The Anatomy of *Paramecium aurelia*

Dr. A. Jurand was a lecturer at the University of Cracow in Poland from 1947 until 1957. Since 1957, Dr. Jurand has been at the Institute of Animal Genetics at the University of Edinburgh where he is a lecturer in Biology.

Dr. G. Selman who also works at the Institute of Animal Genetics has been a lecturer in Genetics at the University of Edinburgh since 1953. Previously Dr. Selman was a demonstrator in the Department of Physics at King's College, University of London.

The Anatomy of

Paramecium aurelia

A. JURAND AND G. G. SELMAN

Department of Animal Genetics, University of Edinburgh

Macmillan

St. Martin's Press

First published 1969

Published by
MACMILLAN AND CO LTD
Little Essex Street London W C 2
and also at Bombay Calcutta and Madras
Macmillan South Africa (Publishers) Pty Ltd Johannesburg
The Macmillan Company of Australia Pty Ltd Melbourne
The Macmillan Company of Canada Ltd Toronto
St Martin's Press Inc New York
Gill and Macmillan Ltd Dublin

Printed in Great Britain by
ALDEN & MOWBRAY LTD
at the Alden Press, Oxford

Foreword

The ciliate protozoan *Paramecium* was formerly much used as an experimental organism by different kinds of biologists. Its study gave rise to heated arguments about the nature of organisms and cells, and to philosophical reflections about immortality, rejuvenescence and so on. More recently – during the past twenty-five years or so – one species, *Paramecium aurelia*, has been popular for certain types of genetical research, especially research concerned with cell heredity and the rôle of the cytoplasm.

In spite of a vast literature, however, and leaving aside a number of older monographs, we have at present no comprehensive account of the morphology of the organism, especially those aspects of it which have been revealed by use of the electron microscope. Consequently the present work by Dr Jurand and Dr Selman is very welcome. It includes many original electron micrographs, of great aesthetic as well as scientific value, and a concise but substantial textual description.

Study of these pictures gives one an impression of the extraordinary structural complexity of the ciliate 'cell' – a complexity which would be incredible were it not for the fact that we have the evidence before our eyes. All of these fibrils, membranes and corpuscles, have to be formed in every developing cell, and hence the material is a marvellous illustration of the phenomena of biological replication and intracellular differentiation. The work is a stimulus to further research into these little understood but vitally important areas of biology.

To the student, perhaps, the book's main value is to show the power of the electron microscope in revealing details of the structures of cells. Above all this is a work of reference which will be essential for all biologists who make use of ciliates in their research or teaching.

Edinburgh, 27 August 1968   G. H. BEALE, F.R.S.

Contents

Plates

Preface

The present work is intended as a detailed illustrated description of the ultrastructure of the protozoan, *Paramecium aurelia*. The photographic illustrations, which consist of light and electron micrographs, represent original observations made (with the exception of four figures) in the Institute of Animal Genetics, University of Edinburgh, during the past few years. Our aim has been to help both student and teacher by supplementing the less detailed accounts based on light microscopy which are available in many textbooks of zoology. *P. aurelia* is a species of considerable scientific importance and this work has been designed to assist protozoologists and those who are not primarily interested in morphology but who may wish to have a convenient source of information about *Paramecium* morphology to help them interpret the considerable volume of scientific literature dealing with the genus.

Most of the descriptions of cell structure and ultrastructure in this volume are appropriate to a paramecium maintained under normal conditions of culture and observed during the period of relatively stable morphology between two successive binary fissions. In this respect our knowledge of the anatomy of *Paramecium* at the ultrastructural level now seems to be fairly complete, so that subsequent advances will be limited to matters of detail. Very little dedifferentiation occurs during any stage in the asexual life cycle of *Paramecium* and therefore most of the ultrastructures will be present all the time. On the other hand there are many features of the fission process, oral development and of conjugation which have not yet been properly studied at the ultrastructural level. Electron microscopy has not yet contributed as much to our knowledge of the structural changes which occur within protozoan nuclei as has the light microscope. But significant observations have recently been made with the electron microscope which bear upon the important problem of how individual cell organelles, such as cilia and trichocysts, are reproduced during the fission cycle and such results will be included.

We have tried to follow the definitions and terminology of Corliss (1959), except that the term 'gullet' is used instead of 'buccal apparatus' as a general term which refers to all the mouth parts; 'cytopyge' has been used for 'cytoproct'. Corliss (1959) and Grimstone (1961) have used the term 'basal granule' instead of its well-known synonym 'kinetosome'. Anterior is defined as the direction towards which a paramecium normally swims and the ventral surface is that in which the gullet is located. The left- and right-hand side of a paramecium is fixed to conform with these concepts of anterior and ventral. In description of structures at the cell surface as seen in tangential section, there are differences which depend upon whether the structures are viewed from the standpoint of an observer within the paramecium looking outwards or an observer outside the paramecium looking inwards. It has seemed natural to us to adopt the standpoint of an observer viewing the surface from outside. Readers should note, however, that in the review of Pitelka (1967) the opposite convention has been chosen. Pitelka's convention has the advantage that it permits agreement with the 'rule of desmodexy' which was enunciated by Chatton and Lwoff (1935), and states that in most ciliates kinetodesmal fibres run anteriorly and to the right of their basal granules. The trouble with this otherwise useful rule is that it depends upon rather a special definition of what is meant by 'right'. You must either, like Pitelka (1967), imagine the observer inside the ciliate and turning so that he always faces the surface under consideration or, like Preer (1967), define 'right' as clockwise to the observer when he looks at the anterior end of the ciliate in polar view.

In the bibliography are listed all those papers known to us which report results obtained by electron microscopy on *Paramecium*, and we apologise to any author whose work may have been accidentally overlooked. Reviews which deal with the ultrastructure of protozoa have also been recorded. In the text we have tried to give a straightforward account of structural and ultrastructural features but every effort has also been made to relate this to

xii

other kinds of work on protozoa by quoting the appropriate references.

In this account we have followed the example of many distinguished morphologists and have in many instances attempted to relate the observed structure of cell organelles with their probable physiological function. Some difficulties arise here, however, because in *Paramecium* the function of many organelles is incompletely understood. Moreover scientific authors frequently find it necessary to warn each other against faulty speculations deduced from structural appearances. Nevertheless we consider that morphologists should use their observations as a basis from which to make inferences concerning the possible function of the organelles they are studying. In this way they may hope to encourage cell physiologists to conduct the necessary crucial experiments.

We are greatly indebted to Dr K. W. Jones of this department for allowing us to reproduce certain of his light microphotographs to illustrate some features of conjugation in *P. aurelia* (Plates 44 and 45). We are grateful to Dr Margaret R. Mott for permission to reproduce Plate 7, fig. 2, which has been published already (Mott, 1965). We also thank Professors T. F. Anderson and J. R. Preer for Plate 13, figs. 4 and 6; Mr R. E. Sinden for Plate 29 fig. 2; Miss Margaret M. Perry and Mr R. E. Sinden for Plate 29, figs. 4 and 5 and Professor J. R. Preer for Plate 51, figs. 1, 2, 3 and 7. Our special thanks are due to Mr E. D. Roberts of our drawing staff for the black and white diagrammatic illustrations. It is a pleasure to acknowledge the encouragement and interest that Professor Waddington has shown us and the valuable advice we have received in discussion with Professor G. H. Beale.

A. JURAND
G. G. SELMAN

1 Introduction to *Paramecium* as an object of scientific study

Paramecium is an animal which is more or less familiar to all who have studied Biology. Scarcely a textbook of Zoology fails to describe this genus, either as an example of a holotrichous ciliate, or as an example of elaborate organisation and specialisation in a unicellular animal. On the other hand most textbook accounts of *Paramecium* give no hint of its ultrastructural complexity as revealed in recent years with the electron microscope.

Despite its long history of scientific investigation many biologists still find *Paramecium* suitable for the investigation of such diverse biological problems as the structure and replication of sub-cellular organelles, cell locomotion by ciliary activity, the mechanism of genic expression and the maintenance of intracellular symbionts.

Paramecium has a world-wide distribution in pools of fresh water which contain decaying organic matter. Paramecia, therefore, must have been among the little animals observed in samples of water by the earliest naturalists to use magnifying lenses in the seventeenth century. For example Dobell (1932), in his account of the life and work of Antony van Leewenhoek, refers to letters and sketches which passed between Christian and Constantijn Huygens and between Leewenhoek and Constantijn Huygens, which show that these observers were familiar with paramecium in 1678. The name *Paramecium* was given by John Hill in 1752 although this pre-Linnean name was not used in the customary binomial manner until the description of *P. aurelia* was given by Muller (1786). Following the introduction and then the progressive improvement of achromatic light microscopes during the nineteenth century and the development of improved methods of fixation and staining towards the end of the century, there was a great increase in the number of scientific papers dealing with the biology of

Paramecium. These papers were not all concerned with structure or comparative morphology, but included valuable work concerned with physiological problems such as the coordinated beating of the cilia and their rôle in locomotion and feeding. Also studied were the movement of food vacuoles, the osmo-regulatory function of the contractile vacuoles and the avoiding reaction which is stimulated by a paramecium meeting an obstacle or a less favourable environment. This kind of work has been reviewed by Kalmus (1931), Dembowski (1962) and Wichterman (1953). The monograph of Kalmus (1931) quotes 833 papers on *Paramecium*; Wichterman (1953) lists nearly two thousand, and so one may estimate that by now there must have been published more than three thousand scientific papers describing work on this protozoan.

A great deal of the modern interest in *Paramecium* has stemmed from the discovery of the system of mating types in *P. aurelia* by Sonneborn (1938b and c) which made it possible to cross stocks of animals which differ genetically. *Paramecium* has other advantages as an animal for laboratory experimentation. Not only may large populations of *Paramecium* be maintained for indefinite periods in small volumes of water under laboratory conditions, but certain features of the life cycle have proved to be especially favourable for the study of both cytoplasmic inheritance and nuclear-cytoplasmic relationships. At conjugation in *P. aurelia*, the two micronuclei each pass through meiosis in both partners and one of the products of meiosis then undergoes a further division to produce two haploid gamete nuclei in each conjugant. One gamete nucleus remains stationary, while the other migrates into its pairing partner and there fuses with the stationary gamete nucleus, so that reciprocal cross-fertilisation occurs and gives rise to genetic recombination. Normally the cytoplasm does not mix during conjugation. In certain environmental circumstances, which are under experimental control, both nuclei and cytoplasm are exchanged. By comparing the results of conjugation with cytoplasmic exchange and conjugation without cytoplasmic exchange, it is possible to distinguish between the effects of cytoplasmic and nuclear determinants. When replicate crosses are made between stocks of animals of different genetic constitution there is little difficulty in obtaining sufficient similar material, as any paramecium can be isolated and allowed to reproduce itself by binary fission to form a clone of individuals of identical genetic constitution. Furthermore, under suitable conditions, autogamy may be induced. In this process (Diller, 1936) the nuclear changes are the same as those associated with sexual reproduction which normally follows conjugation, but in autogamy no pairing takes place and self-fertilisation occurs between the two haploid gamete nuclei of the same animal. Since both these nuclei were derived from the same haploid product of meiosis, the resulting individual may be allowed to develop into a clone of individuals which are completely homozygous in respect of all their genes. The behaviour of the nuclei during the various reproductive stages has been described by Sonneborn (1947) and an account of the way in which the advantages of the material are being exploited for the purposes of genetic analysis is given by Beale (1954) and Preer (1967).

From the cytological point of view there is no doubt that each individual paramecium is a single cell, for it is bounded by a single continuous plasma membrane. On the other hand, when its animal behaviour is considered, *Paramecium* shows a sophistication which invites comparison with some orders of metazoa. Diverse functions in *Paramecium* are performed with the aid of organelles which are specially differentiated and are made up of macromolecular complexes which are organised at a sub-cellular level within particular regions of the cytoplasm. It is this kind of organisation which can be well demonstrated by electron microscopy. Yet it must be stressed that differentiation in the metazoa is organised on quite a different kind of plan. In the higher animals, bodily functions such as digestion, excretion, osmoregulation and locomotion are performed by particular organs made up of assemblies of different kinds of interdependent differentiated cells; and each cell is a living unit which has become specialised during embryonic development to perform one particular function or limited group of functions. The ultrastructure of each differentiated cell, therefore, is of a particular recognisable type by which the function of that cell may be inferred.

One of the obstacles in the way of studying the nature of differentiation in metazoa lies in the fact that differentiated cells outside the germ-line cannot be induced to form germ cells, so that a genetic analysis of such cells is impossible. In *Paramecium* on the other hand this difficulty does not exist because cross matings are made between fully differentiated unicellular animals. *Paramecium* in fact may be regarded as both a germ cell and a somatic cell. Its phenotype has been shown to be under the control of the single compound macronucleus (Sonneborn, 1947) while either of the two micronuclei may provide the gamete nuclei. Both the macronucleus and the micronuclei of an individual are in normal circumstances derived from the single diploid zygote nucleus which was formed after the previous conjugation or autogamy.

Genetic researches into the antigen system of *P. aurelia* have led to a concept of the existence of alternative and mutually exclusive cytoplasmic states, a number of which may occur within any given stock, and which may be evoked by the application of appropriate environmental stimuli. Also, different genes may be expressed in different cytoplasmic states. These ideas of Sonneborn (1947) and Beale (1954) offer a model for cell differentiation but further research is needed before it can be decided whether the model is applicable outside the protozoa.

2 Introduction to electron-microscope studies with *Paramecium*

All the main organelles of *Paramecium* have been carefully observed and described by light microscopists such as J. von Gelei (1934), G. von Gelei (1937) and Lund (1933, 1941), but such work was always hindered by the limits of optical resolution. The electron microscope gives a resolving power one hundred times greater than does the light microscope and fine detail of 2 to 3 mμ may now be distinguished in thin sections of material. This means that macromolecules may be observed as well as structures made up of the ordered arrangements of macromolecules. As a result, a wealth of unsuspected ultrastructural detail has been revealed during the past eight years in every cell organelle of *Paramecium*. It may be noted that the technical advance which finally opened the way to this new information was not the introduction of the electron microscope but the introduction of successful ultra-thin sectioning techniques by Porter and Blum (1953) which enabled the new instrument to be applied to the study of structure inside the cell. These advances do not mean that the light microscope has been superseded, for observations on live specimens may only be made by light microscopy and in certain work with fixed and stained specimens, the light microscope will be chosen for convenience and to save time.

Most papers which report electron-microscopic observations made on *Paramecium* concern themselves with a particular organelle or group of structures. Although there are about twelve known species of *Paramecium*, ultrastructural studies have been made on only four of them. Observations with the electron microscope have been made on *P. aurelia* by Dippel (1964, 1965), Ehret and Powers (1959), Jurand (1961), Jurand, Beale and Young (1962, 1964), Mott (1963, 1965), Schneider (1959a and b, 1960a and b, 1963, 1964a and 1964b) and Stewart and Muir (1963). All the diagrams and photographs in the present work are of *P. aurelia*. *P. multimicronucleatum* was used in the work of Inaba, Suganuma and Imamoto (1966), Mijake (1966) and Pitelka (1963, 1964, 1965, 1967). *P. caudatum* has been studied by Wohlfarth-Bottermann (1956a, 1956b, 1957, 1958a), André and Vivier (1962), Vivier and André (1961a and 1961b), Schneider (1959a and b, 1960a) and Yusa (1963, 1964). *P. bursaria* was used by Ehret and Powers (1957, 1959), Ehret and de Haller (1963) and Ehret, Savage and Alblinger (1964). In addition Grimstone (1961) and Pitelka (1963, 1967) have published useful reviews of certain aspects of ultrastructure in the ciliates and in other classes of Protozoa. No previous attempt has been made, however, to provide a complete description of the morphology of *Paramecium* since the introduction of the electron microscope. Indeed Pitelka (1963) remarked that there is not a single protozoan type whose ultrastructure and ultrafunction has been as thoroughly mapped as have some of the more familiar types of vertebrate cells. We feel that of all protozoans, *P. aurelia*, by virtue of its scientific importance most deserves to have its ultrastructure comprehensively described. The relevant observational information is available for *P. aurelia* although it is widely distributed in the scientific literature.

The majority of electron micrographs used as illustrations in the present work have been made from material prepared in a manner which is basically that used by most electron microscopists. The method in outline consists of fixation in osmium tetroxide solution, dehydration in ethanol, embedding in plastic and sectioning at a thickness of about 80 mμ. Details are given in Appendix I. The difficulties involved in the interpretation of electron micrographs are similar to those encountered in the interpretation of fixed and sectioned material observed by light microscopy but in practice the higher magnifications used in electron microscopy compel the use of far greater care if serious artefacts are to be avoided. Not only is there the possibility of introducing artefacts during fixation and embedding procedures but maladjustments of the electron optical system can easily introduce further artefacts. Fortunately the present state of electron microscopy is such that the more obvious pitfalls may be recognised and avoided, so that the non-specialist may for the most part have reasonable confidence that the structures which are interpreted by the electron microscopist do correspond

6

to similar structures present in the living cell. A full discussion of this question would be out of place here but some of these matters have been considered for example by Agar (1965), Harris (1962) and Pease (1960).

It is considered a good practice in light microscopy to compare the results obtained by fixation using a variety of methods. The situation in electron microscopy is that a variety of methods of fixation (such as the use of formaldehyde, glutaraldehyde, permanganate, freeze-drying and glutaraldehyde followed by osmium tetroxide) have been tried by different authors on various materials; although the results obtained with these different methods do not seriously conflict with one another, the general conclusion may be drawn that at present there seems to be little to be gained from the use by the morphologist of known fixatives other than osmium tetroxide or glutaraldehyde followed by osmium tetroxide. There is good evidence that the use of osmium tetroxide allows reliable fixation of cell structures which are composed of membranes or layers of lipoprotein or which consist largely of fibrous protein. The use of glutaraldehyde enables certain fine elements of ultrastructure such as microtubules to be preserved, but most of these structures can also be observed without the use of glutaraldehyde if there is calcium in the buffered solution in which the osmium tetroxide is dissolved.

The study by electron microscopy of developmental processes such as fission and conjugation presents special difficulties. Not only must the electron microscopist know the orientation and exact location of his field of view but he must also know the exact stage at which the paramecium was fixed, so that comparisons may be made between similar fields of view in different animals. Unfortunately development does not take place synchronously even within a clone of animals under standard conditions, so that the paramecia must either be observed, timed and fixed individually or a selection must be made after fixation with reference to some 'yardstick' of development such as the observed state of the nuclei during conjugation or the length of the animal during fission (Ehret and de Haller, 1963). Such procedures take a long time and the intracellular detail is less well preserved after individual fixations have been made, perhaps because of the increased handling to which the material must be subjected. Furthermore if it is

7

suspected that a particular organelle is increasing or decreasing in number or in size as development takes place, then many observations will need to be made on different animals within the same clone and it will be necessary to make a good estimate of the fixation time of each animal observed. These are some of the reasons why the study of development at the ultra-structural level in *Paramecium* is now at a less advanced stage than the study of the more static features of its anatomy.

Two further warnings may be worth offering. Firstly, it is necessary to be rather sceptical about estimates of size made by different electron micro-scopists, for variations in preparation techniques may introduce different degrees of shrinkage or swelling. Thus if one author estimates that a particu-lar kind of granule is 7 mμ in diameter and another estimates it to be 10 mμ, they may still be observing the same kind of granule. Secondly, it is necessary to be particularly wary of dynamic interpretations based upon a series of electron micrographs unless it is clear that the observer has made his fixations against an appropriate developmental time-scale.

Electron microscopy should never be regarded merely as a tool for the observation of cytological detail. Its use may be profitably combined with other techniques such as histochemistry, autoradiography and immunology to study a wide range of problems in cell biology (Haggis, 1966). For instance a great deal of work has been done on the genetic analysis of the variation in surface antigens in *P. aurelia* (Beale, 1954; Preer, 1967). Treat-ment of whole paramecia with ferritin-labelled antibody followed by thin sectioning enabled Mott (1963, 1965) to demonstrate by electron microscopy the presence of these antigens on the surface of the plasma membrane both at the cell surface and on the surface of the cilia (Plate 7, fig. 2). During the transformation with change of temperature, from one antigenic type to another, the new antigen appeared first at isolated sites on the cell surface and then spread to the cilia until the whole of the plasma membrane with the exception of the gullet area was covered.

Several accounts of ultrastructure within the Protozoa have been written which also give cautious consideration to some of the problems of com-parative anatomy and systematics (Grimstone, 1961). Examination of the ultrastructure of organelles and components of *Paramecium*, demonstrated both similarities and differences between the structures in *Paramecium* and those in cell organelles which have a similar function in other groups of protozoa and in the cells of metazoa. When similarities are found, it is worth remembering that not only can these be cited as evidence in favour of a common evolutionary origin, but they may equally well be taken as evidence of evolutionary convergence. The latter concept arises because certain structures, such as the basic nine plus two microfibrillar structure which is found in motile cilia from so many forms of life, may represent a superior solution to a basic problem which has been arrived at in the many forms because of its superiority. Microscopical observations have frequently been used as a starting-point for speculations about evolutionary origins but unfortunately in this field there have been examples of conflicting opinions each of which was derived from a consideration of the same data. It would seem that in our present state of knowledge such speculations have little scientific value.

3 The general morphology of *Paramecium* observed by light microscopy

Paramecium is an elongated cell with an outline not unlike that of a slipper (Plate 1). In this analogy the toe of the slipper corresponds to the posterior end and its heel corresponds to the animal's anterior end. Thus the posterior half of a paramecium is slightly wider than its anterior half; and the posterior end of *Paramecium* is bluntly pointed while the anterior end is rounded. Upon meeting an obstacle when swimming, *Paramecium* reverses its direction, travels backwards for a short distance, pivots and then resumes its normal forward motion upon a different course. This behaviour is called its typical 'avoiding reaction', as described by Jennings (1906). Dembowski (1924) and Golinska (1963) have also described how *Paramecium* may rebound from an obstacle without any reversal of the forward swimming.

In transverse section *Paramecium* is oval in outline (Plate 14) with dorsal–ventral flattening. A shallow groove, called the oral groove (Plate 1 and Plate 3, fig. 1), runs from the anterior end along the ventral surface to a point about midway between the anterior and posterior poles. The oral groove is not quite straight but imparts a slight dextral spiralization to the anterior half of the animal. On the other hand, Parducz (1962) has shown that when swimming forward normally *Paramecium* (except for *P. calkinsi*) spirals through the water in a sinestral path. Parducz goes on to point out that although the dextral spiralisation of the groove must therefore be a hindrance to efficient swimming, its shape helps to collect food in the gullet. It will be obvious that *Paramecium* possesses no plane of symmetry.

P. aurelia is on average 135 µ long and 40–45 µ across its widest diameter. The dimensions, however, of any individual specimen will vary according to its genotype, its conditions of culture and the time with respect to its fission cycle at which it is observed. Under comparable conditions there are

10

considerable differences in length and breadth among different syngens (see Chapter 11). For example the length varies from a mean of 112 μ in syngens 4, 8 and 10 to 204 μ in syngen 12 (Sonneborn, 1957).

Certain prominent organelles of *P. aurelia* may be clearly seen with the light microscope (Plate 1). These include the gullet, food vacuoles, the macronucleus and micronuclei, and the contractile vacuoles. Electron microscopy is, however, required to observe their structure in detail.

At the posterior end of the oral groove, in the middle of the animal's ventral surface, there is a depression called the vestibulum which leads into a funnel-like gullet. The gullet or buccal cavity extends posteriorly and inwards until it terminates at a specially differentiated part of the buccal wall called the cytostome below which food vacuoles are formed. The beating of the cilia in the oral groove, vestibulum and gullet drives bacteria into the bottom of the buccal cavity.

The food vacuoles can be seen moving about in the interior of the animal while the bacteria are progressively digested. The unassimilated remnants are excreted from the anal pore of the cell (also called the cytoproct or cytopyge) which has a fixed position on the ventral surface, posterior to the gullet (Plate 1). Two contractile vacuoles may be observed on the dorsal surface, one in the anterior and the other in the posterior half of the animal. *P. aurelia* normally has one macronucleus and two micronuclei which may be observed in fixed preparations when, for example, basic dyes or the Feulgen technique are used. The outer surface or pellicle of a paramecium is covered with cilia, which normally beat so fast that only a faint flickering may be seen at the edge of the animal. At the posterior pole there is a tuft of immotile longer cilia (Plate 11, fig. 4). Light microscopy of fixed preparations stained by techniques such as those of Klein (1926), von Gelei (1937, 1939) and Chatton and Lwoff (1936) show that each fully formed cilium is associated with one unit of a complicated but regular silver-line pattern which repeats itself over the entire pellicle (Plate 2). Not all the silver techniques give exactly the same results and the interpretation of the silver-line images gave rise to considerable controversy which continued until the surface structures of *Paramecium* had been studied by electron microscopy. One of the most useful aspects of the review of Ehret and Powers (1959) was in fact the achievement of a plausible reconciliation between the results of light microscopists and the new observations made by electron microscopy. Dippell (1962) investigated the sites of the silver deposition by both light microscopy and electron microscopy, and showed that silver grains were deposited throughout the parasomal sac and just above both basal granules and trichocysts. Recently Gillies and Hanson (1967) and Kaneda and Hanson (1967) have taken advantage of the fact that silver staining can be used to observe the positions of these organelles to study morphogenesis by light microscopy during the fission cycle of *P. trichium* and *P. aurelia*.

4 The ultrastructure of the pellicle and its associated organelles

The cell cortex of *Paramecium* is made up basically of similar corpuscular units (Plate 4) arranged in longitudinal rows (Plates 5 and 6). On the dorsal surface the units may be arranged in hexagonal close packing (Plate 3, fig. 4) so that, to use the analogy of Ehret and Powers (1959), they resemble a single layer of soft peas packed in the bottom of a bowl. Pitelka (1967) employs the term 'kinetosomal territory' in the same way as the term 'corpuscular unit' is used by Ehret and Powers (1959) and Ehret (1960).

A motile cilium, or frequently a pair of cilia, emerges from a boat-shaped depression in the cell surface at the centre of each corpuscular unit while beneath each cilium lies the basal granule (Plate 7 and Diagram 1). The shape of the corpuscular unit is determined by a pair of alveoli whose membranes are opposed to form a median septum (Diagram 2) along a line which runs longitudinally through the base of each cilium. The alveoli lie along the sides of cytoplasmic ridges which run longitudinally midway between the rows of cilia (Diagram 3).

At the surface, the pellicle is covered by a single unit membrane which is ridged and furrowed as it follows the shape of each corpuscular unit, and which is also continuous with the membrane which covers each cilium (Plate 7). If, however, a needle be imagined to penetrate the cell, it would pass through three unit membranes and a granular layer (Plate 8), unless the position of entry were above the median septum or the border between adjacent corpuscular units. The surface membrane lies immediately above the outer alveolar membrane, so that the alveoli are separated from the outside by two similar unit membranes (Plate 8). Each alveolus is bounded by a membrane which on its upper surface is called the outer alveolar

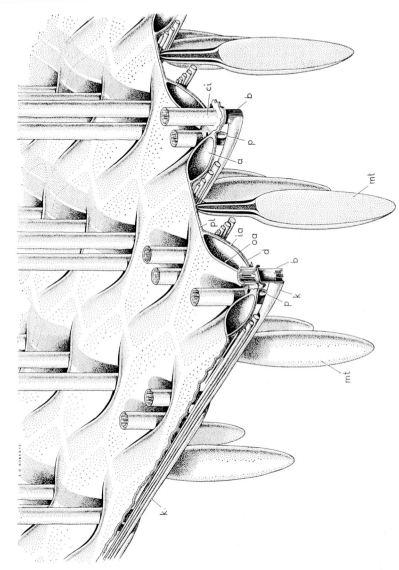

DIAGRAM I   This three-dimensional drawing of part of the pellicle of *P. aurelia* demonstrates the spatial relationship between the organelles which are shown in plan view in Diagrams 2 and 3. The manner in which individual kinetodesmal fibrils (k) overlap within a kinetodesma is shown along the left edge of the diagram. The kinetodesmata run below ridges in the pellicle between adjacent kineties and along each kinety the cilia (ci), or in some cases pairs of cilia, alternate with trichocysts (mt). Also present are basal granules (b), parasomal sacs (p), the dense granular cortical layer (d), alveoli (a), the plasma membrane (pl), the outer alveolar membrane (oa), and the inner alveolar membrane (ia). Microtubular ribbons and fine fibrils are omitted from this diagram but they are shown in Diagram 3 and in Plate 10.

membrane and on its lower surface is called the inner alveolar membrane. The inner alveolar membrane lies immediately above the granular layer (Plate 8, fig. 2) which is about 12 mμ thick.

DIAGRAM 2   This diagram of the pellicle in plan view shows the relative positions of cilia (ci), the parasomal sac (p), trichocyst tips (tt), alveoli (a) and kinetodesmal fibrils (k). The two alveoli (a), on either side of the median septum (s), together have a hexagonal outline which indicates the limits of one corpuscular unit. The two cilia are shown as they would appear in a section cut tangentially to the cell surface (as in Plate 10, fig. 1). They are shown immediately surrounded by a boat-shaped circumciliary space between the alveoli. The circumciliary space is outside the surface of the paramecium and is separated from the alveoli by the plasma membrane and the outer alveolar membrane, each of which is here represented by a single line. A basal granule lies immediately beneath each sectioned cilium shown here, and a kinetodesmal fibril joins the posterior basal granule of the pair. In some cases there is continuity, below the surface ridges, between alveoli in adjacent corpuscular units but never across the median septum.

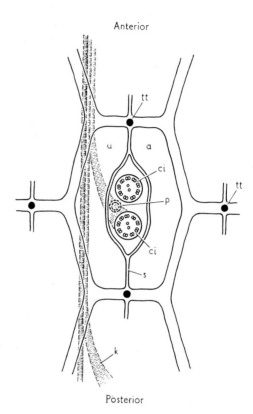

Anterior

Posterior

The cell membrane at the surface of *Paramecium* and both the inner and outer alveolar membranes all show the typical three-ply unit membrane structure of Robertson (1959, 1960) which has been observed, by electron microscopy, in cell membranes throughout the animal and plant kingdoms. Each unit membrane consists of two dark osmiophilic layers which are 2 to 3 mμ thick and are separated by a light space of thickness 3 mμ (Sjöstrand and Elfin, 1962). The outer alveolar membrane generally runs beneath the cell membrane, separated from it by about 10 mμ but occasionally this distance is increased to accommodate a mass of frothy vesicular cytoplasm (Plate 9, fig. 1). This vesicular material is most abundant near the median septum.

Each corpuscular unit is seldom a regular hexagon in plan view as the opposite internal angles pointing in the lateral directions are usually greater than 120° (Diagram 2) and often approach 180° when the pattern consists of longitudinal rows of rectangular units. Along each such row, exactly midway between adjacent kinetosomes and just below the cell surface where adjacent alveolar membranes almost meet, are the tips of remarkable subcortical flask-shaped structures called trichocysts (Plates 4 and 12). Immediately under the longitudinal ridges between adjacent rows of cilia run bundles of fibres called kinetodesmata (Plates 9 and 10). Each kinetodesma consists of a number of kinetodesmal fibrils each of which originates at a basal granule (Plate 9, fig. 3) beneath a motile cilium and then runs diagonally to the ciliate's left (viewed from outside) and anteriorily to join the bundle of fibrils in which it runs for a distance equal to the length of four or five corpuscular units (Diagram 3). Just anterior to the basal granule with the kinetodesmal fibril attached to it, there may be a second basal granule in the same corpuscular unit, but without a kinetodesmal fibril (Plate 11, figs. 5 and 9).

Immediately to the left of each basal granule or pair of basal granules there is a small tube-like structure called the parasomal sac (Diagram 2). The sac appears to connect with the cell surface via a tube about 80 mμ in diameter, the wall of which is thickened to form a ring as it passes through the granular layer just below the level of the inner alveolar membrane. Hufnagel (1967) has pointed out that even in a tangential section at the level of the alveolus the sac is seen to be connected with the posteriormost basal granule by a thin sheet of membrane-bounded cytoplasm (Plate 11, fig. 8). Hufnagel (1966) also reported further ultrastructure in the parasomal sac, including five notches on the outer side of the ring remote from the basal granule and some fine serrations on the inner margin of the ring.

It must be emphasised that each of the structures or cell organelles mentioned above occupies a definite fixed position with respect to the others so that the entire pellicular system, in plan view (Diagram 3), has geometrical properties such that it could be used as a repeating wallpaper design with the pattern continued on adjacent rolls. In this analogy, if each roll of wallpaper were printed so as to show one longitudinal row of kinetosomes down its mid-line, then each roll could be said to include one 'kinety' which Corliss (1959), following Chatton and Lwoff (1935), defined as a line of kinetosomes with its associated kinetodesmal fibrils.

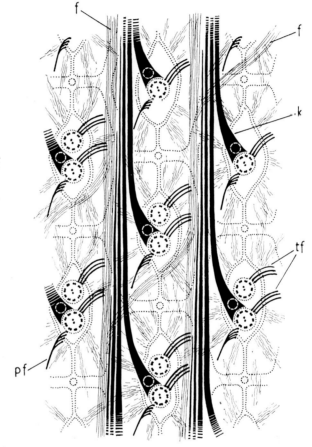

DIAGRAM 3 A diagram of the pellicle in plan view to show the infraciliature which consists of kinetodesmal fibrils (k), the transverse tubular fibrils (tf), the postciliary tubular fibrils (pf) and the network of fine fibrils (f) which are represented here by fine lines. These elements of infraciliature have been drawn superimposed upon a plan view of the pellicle (as in Diagram 2) which appears as dotted lines. A comparison may also be made with Plates 4 and 10.

This analogy describes fairly exactly the morphology of the pellicle over most of the dorsal surface of *Paramecium*. But it also emphasises a difficulty, in that it is hard to understand how such a regular pattern could be maintained over the three-dimensional non-symmetrical surface of the *Paramecium* cell. The difficulty in maintaining the pattern would appear to be greatest at the two ends of the animal and on the ventral surface where there are irregularities imposed by the presence of the oral groove and the vestibular depression which leads into the gullet. In these regions the difficulty is met in part by a failure to maintain the corpuscular unit as a regular hexagon or rectangle, as described above and allowing it to exhibit a deformed, almost rhombic shape. In addition, along the mid-ventral surface of *Paramecium* there is a line of suture (Plate 2, fig. 1 and Plate 3, fig. 2) which runs forwards from the gullet to the anterior end and back from the gullet to the cytopyge (Plate 2, fig. 1) and thence to the posterior end. At the pre-oral suture, two rows of cilia meet at an angle of about 90°. They meet at an angle of about 30° however at the post-oral suture. Pitelka (1965) illustrates the ultrastructure at the pre-oral suture and shows that the kinetodesmata actually cross each other along the line of suture.

The pellicle system also shows certain other departures from strict geometrical regularity. For instance, although there are two separate alveoli in each corpuscular unit, connections are commonly found between alveoli in adjacent units of the same longitudinal row (Diagram 2, Plate 6 and Plate 10, fig. 1). These connections may occur to either side of the tip of a trichocyst between alveoli on the same side of the median septum; and electron micrographs show that a string of four adjacent units may be so connected to each other along a longitudinal row. Connections have also been observed between alveoli, across the cytoplasmic ridges which run between adjacent longitudinal rows. On the other hand all electron micrographs show that the median septum, which runs longitudinally to separate the alveoli to either side of each corpuscular unit, is always maintained intact.

It should also be emphasised that although the tips of the trichocysts always occupy their fixed position in the pattern of the cortical pellicle, not all trichocyst sites are occupied. We have examined patches of twenty-five adjacent trichocyst sites in electron micrographs and found that the number occupied varied between seven and seventeen.

THE CILIA AND THE BASAL GRANULES

The cilia and basal granules of *Paramecium* are of a conventional ultrastructure (Diagram 4 and Plate 7) which does not differ basically from that

18

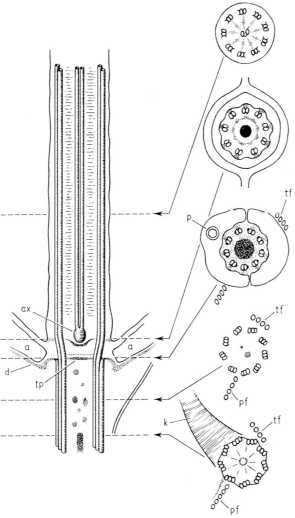

DIAGRAM 4 This is a diagram of a cilium and its basal granule in longitudinal section, together with five transverse sections at levels which are indicated. The ultrastructures are based upon Plates 7 and 11, and the observations of Pitelka (1965) and other authors. The transverse sections are drawn from the point of view of an observer outside the paramecium, with the anterior direction towards the top of the page. There is continuity between the paired peripheral microtubules which run beneath the membrane of the cilium and two microtubules of the triplets of three which outline the basal granule. The two axial microtubules of the cilium terminate at an axial granule (ax). Also present are the alveoli (a), cortical layer (d), kinetodesmal fibril (k), parasomal sac (p), transverse tubular fibrils (tf), postciliary tubular fibrils (pf), and the terminal plate (tp).

in other protozoa or higher animals. The structures have previously been described for *P. multimicronucleatum* by Pitelka and Child (1964) who also summarised what is known about the problems of locomotion by the coordinated beating of cilia, in which field of study *Paramecium* often has been used as an experimental animal.

In ciliates the basal granule or kinetosome is regarded as a centre of organisation for morphogenesis (e.g. see Lwoff, 1950) within its own territory but how it may operate as such at the macromolecular level is quite unknown (Pitelka, 1967). A later chapter will give an account of electron-microscope observations which have shown that in synchrony with the fission cycle, kinetodesmal fibrils, microtubular ribbons and alveoli all take form in a fixed relationship to a basal granule which itself has previously formed in front of a pre-existing basal granule. Studies of growth and cell duplication in *Paramecium* demonstrate that each cilium develops by rapid growth from an underlying basal granule, although so far there are no high resolution electron micrographs to show the stages of the development. Pitelka and Parducz (1962) showed that cilia may be stripped from *P. multimicronucleatum* by treatment with nickel ions and the cilia may subsequently be regenerated from the basal granules. Deciliation can also be induced with chloral hydrate (Alverdes, 1922; Grebecki and Kuźnicki, 1961; Kennedy and Brittingham, 1968) after which *P. caudatum* can regenerate its cilia in about 4 h (Kuźnicki, 1963) but it is not known in this case if the basal granules are impaired by the treatment. Observations by Ehret and de Haller (1963) on cilia formed during the normal development of a gullet suggest that cilia can develop from basal granules in as little as 10 min. Smith-Sonneborn and Plaut (1967) have localised DNA at the sites of the basal granules in the pellicle of *P. aurelia*; Randall and Disbrey (1965) found DNA in basal granules of *Tetrahymena*.

Electron microscopy demonstrates that the basal granule in *Paramecium aurelia* is a complex organelle in the form of a hollow cylinder about 0.5 μ in length and about 200 mμ in over-all diameter (Plates 7 and 11) oriented in a direction perpendicular to the cell surface and located immediately below it. The walls of the cylinder are constituted by nine equally spaced fibrils, each fibril being made up of three tubular subfibrils which share a common wall along the edge where they are in contact with each other (Plate 11, fig. 9). Each tubular subfibril of the triplet has a total diameter of about 25 mµ and in cross-section is seen to consist of a dense osmiophilic wall about 8 mµ thick which surrounds a less dense lumen.

Sections through the lower third of a basal granule show a structure which resembles a water-wheel with a hub and nine fibrous spokes which run radially between the hub and the innermost subfibril at one end of each triplet (Plate 11, fig. 7). For each triplet the centres of the subfibrils are in a straight line which diverges from the hub of the wheel in a clockwise direction when viewed from above. In the uppermost two-thirds of the basal granule there is no water-wheel structure but there are some rather ill-defined granules in the lumen of the basal granule. Towards its top, the outermost subfibril of each triplet tapers and terminates.

The top edge of the basal granule is also marked by a dense plate (Diagram 4 and Plate 7), about 20 mµ thick, across the lumen at the same level as the dense fibrogranular layer under the inner membrane of the alveoli. Hufnagel (1966) showed that it is possible to isolate this plate as a separate disc with nine perforations through which pairs of microtubules normally pass. There is also another thin transverse layer, 5 mµ thick, just above the thick layer at a separation of 3 mµ. By contrast there is no terminal plate at the lower edge of the basal granule.

Where the basal granule bears a cilium, the innermost pair of subfibrils from each triplet of microtubular subfibrils in the basal granule is continuous with the doublets which are equally spaced to form a microtubular frame-work inside the cilium. A cross-section through a cilium (Plate 11, fig. 3) shows that each doublet consists of microtubular subfibrils of similar dimensions to those found as triplets in the basal granule. In addition, at least in the lower part of the cilium, each doublet has a pair of horn-shaped projections which protrude from one edge of each doublet in a clockwise direction, when viewed from outside the cell surface. In cilia from *Tetrahymena* these projections have been shown to possess ATPase activity (Gibbons, 1963) and so they may play an important rôle in the bending movements of cilia.

The cilium shaft has a total average diameter of about 280 mµ and is bounded over all its surface by a unit membrane. Down the centre of the cilium runs a pair of single, separated microtubules which meet together only at the base of the cilium in a dense central knob called the axial granule or axosome (Diagram 4 and Plate 7). Immediately below the axosome there is a curved septum across the cilium and just above the axosome is some rather diffuse material. Electron micrographs of cross-sections of cilia frequently show less clearly defined structures and strands of material which have been described by various authors using different biological material (Gibbons and Grimstone, 1960; Gibbons, 1961; Lansing and Lamy, 1961; Pitelka, 1965; Roth and Shigenaka, 1964; Satir, 1965). There is, however, as yet no general agreement about the form of these less obvious structures. Electron micrographs of cilia often show outward projections or folds in the surface membrane of the cilium. Sometimes slender tubular evaginations or keel-like folds may be present in this membrane. Pitelka (1965) regards all such protuberances as artefacts of preparation and suggests that they reflect a looseness and extensibility of the cell membrane.

THE INFRACILIATURE

A kinetodesmal fibril shows a definite cross-banded structure by which it may be recognised (Plate 9, figs. 2 and 3; Plate 10, fig. 1). The major periodicity is at an interval of about 28 mμ within which interval at least three osmiophilic sub-bands are easily observed and smaller sub-bands are also present. The banded pattern can be seen both in sectioned material and in negatively stained fragments (Pitelka, 1965) and although there is a resemblance to collagen fibres the periodicities seem to be quite different. Hufnagel (1966) found that under adverse conditions the fibril may be dissociated into subfibrils which are 4 mμ wide.

The kinetodesmal fibrils are thickest at their posterior ends where they meet the base of a basal granule. At this end they are irregularly oval in cross-section with a major and minor axial diameter of about 180 mμ and 90 mμ respectively. As they progress anteriorly under the cytoplasmic ridges they taper uniformly throughout their length of about 8 μ. The kinetodesmal fibrils run quite straight and separately from each other in the kinetodesmal bundle where they overlap in the manner of slates on a roof (Metz, Pitelka and Westfall, 1953). In *P. aurelia* the kinetodesmal fibrils do not show relational coiling about each other, as was reported by Ehret and Powers (1959) for *P. bursaria*. Dippell (1964) observed that where there is a pair of cilia at the centre of a corpuscular unit of the pellicle, the kinetodesmal fibril is always found attached to the base of the posterior basal granule of the pair (Plate 11, fig. 9). The anterior basal granule of a pair is never attached to a kinetodesmal fibril.

Just below the level of the kinetodesmal fibrils but at the level of basal granules runs a complicated network consisting of large bundles of fine fibrils, each about 3 mμ in diameter (Plate 5 and Plate 10, figs. 2 and 3). The most conspicuous of these bundles run longitudinally beneath the kinetodesmata (Roth, 1958); there are also rather irregular cross-connecting bundles which run between the longitudinal bundles and resemble them in thickness. These bundles all run parallel to the cell surface. In addition there are smaller, rather whispy, bundles of similar fine fibrils which are observed to run from the proximity of the basal granules outwards and obliquely upwards in all directions until they reach the borders of the individual corpuscular unit (Plate 4).

Certain organised arrays of microtubules have been observed in a fixed relationship to the basal granules (Diagram 3 and Plate 11, figs. 7 and 9). A single line of four or five microtubules originates at the anterior right-hand edge, viewed from outside the paramecium, and tangentially to the base of every basal granule. From here, the ribbon of microtubules (called the transverse tubular fibrils) runs somewhat anteriorly and obliquely upwards to the cell surface of the right-hand ridge. There is also a radial line of four or five microtubules which originates at the left posterior margin of each single kinetosome or of the posterior member of any pair of kinetosomes. This ribbon of microtubules (called the postciliary tubular fibrils) runs posteriorly and obliquely upwards to the cell surface and then towards the left-hand ridge. These microtubules are each about 25 mμ in diameter with a less dense lumen of about 10 mμ diameter. Although these microtubules are well preserved after fixation in 1 per cent osmium tetroxide with sucrose and calcium chloride (0.3 mg per ml), they are not preserved using the same

23

fixative without calcium chloride. In this they are unlike the microtubular elements of cilia and basal granules which therefore appear to possess greater chemical stability.

When the basal granules at the centre of a corpuscular unit are paired, the two members of the pair are joined by an arched bridge of osmiophilic fibrous material, which always arches away from the position of the para-somal sac on the opposite side of the basal granule (Plate 11, fig. 5).

THE TRICHOCYSTS

The trichocyst body is fusiform as is a carrot (Diagram 5 and Plate 12, fig. 1) but in place of the leaf stems, it has a structure which is shaped rather like a thorn and is called the tip of the trichocyst. The trichocyst body is about 5.5 μ long and 1.7 μ across its widest diameter. Trichocysts, being large and numerous (there are about fifteen hundred in *P. aurelia*) occupy a considerable proportion of the cytoplasmic volume of the animal.

The mature trichocyst body is not at all osmiophilic (Plate 12). In these electron micrographs it appears only very slightly darker than the medium outside the paramecium and no structure is discernible within it. This is in marked contrast to the bodies of trichocysts from *P. bursaria*, as described by Ehret and de Haller (1963) which were osmiophilic and showed crystalline structure. This description corresponds to that of Stewart and Muir (1963) who also used *P. aurelia*, and to that of Sedar and Porter (1955) who used *P. multimicronucleatum*.

The trichocyst tip of *P. aurelia* is osmiophilic, about 1.6 μ long, and shows a crystalline structure (Plate 12). The tip frequently shows a periodicity of about 7 mμ between parallel layers of material in planes perpendicular to the long axis of the trichocyst. Similar periodicities are also observed in other micrographs between parallel layers each at an angle of about 60° to these perpendicular planes. Herring-bone patterns have also been observed in sections through the trichocyst tip (Plate 12, fig. 3 and Plate 13, fig. 4) so that our observations for the trichocyst tip show a material similar to that described by Ehret and de Haller (1963) for the whole of the trichocyst. The

24

trichocyst tip of *P. aurelia* seems to be bounded by a triple-layer membrane except where it joins to the trichocyst body.

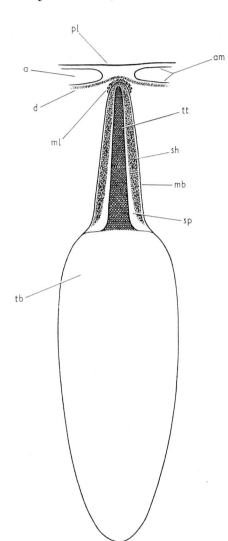

DIAGRAM 5 This is a longitudinal median section through a mature trichocyst in its position below the pellicle. The trichocyst tip (tt) has a crystalline structure with periodicities which appear, in electron micrographs, as three sets of dark parallel lines (see Plate 12, fig. 3, and Plate 43, fig. 2). The sheath (sh) has an irregularly granular structure and the trichocyst body (tb) appears structureless. There is a cluster of microtubules (ml) where the sheath comes into contact with the dense cortical layer (d) of the pellicle which runs beneath the alveoli (a) bounded by alveolar membranes (am) below the plasma membrane (pl). There is a clear space (sp) between the sheath and the tip of the trichocyst. A membrane (mb) runs outside the sheath and also round most of the trichocyst body.

The tip is surrounded by a cap or sheath of dense fibrigranular material about 50 mμ thick (Plate 12, figs. 4 and 6). No dense granules of about 25 mμ diameter have been observed in *P. aurelia* between the trichocyst tip and its sheath, although these were observed in *P. multimicronucleatum* by Sedar and

Porter (1955). There is frequently a clear non-staining gap between the tip and its sheath particularly round the widest part of the tip which is near its junction with the trichocyst body (Plate 12, figs. 3 and 4).

A unit membrane bounds the trichocyst and its sheath to separate it from the surrounding cytoplasm. This membrane has not been observed to pass over the whole of the lower surface of the trichocyst body although it may do so and failure to observe it may be an artefact. This membrane is reinforced, as it passes over the apex of the sheath, by an array of microtubules, each about 25 mμ in diameter. These are located just below the cell surface and are best observed in sections cut tangentially to the tip of the trichocyst (Plate 4).

When a paramecium is irritated by certain stimuli, such as acid in the medium or electrical impulses, the trichocyst tips are discharged into the medium by a remarkable unidirectional expansion of the trichocyst body, which then becomes the trichocyst shaft. After extrusion (see Plate 13), the material of the trichocyst shaft is seen to possess a regular cross-banded structure with a transverse osmiophilic band at intervals of 55 mμ (Jakus, 1945; Jakus and Hall, 1946). There are also some fine fibrils about 3 mμ thick, which are arranged along the length of the shaft (Plate 13, fig. 3). Because of its change in structure on extrusion, the material of the mature trichocyst body must in fact possess a regular ultrastructure which has not been observed.

When an attempt is made to fix paramecia using a slowly penetrating fixative, the surface layer may be affected before the interior cytoplasm. In such cases the mature trichocysts have been observed to discharge themselves so that their shafts expand backwards into the interior cytoplasm, while their tips are held at the surface (Plate 13, fig. 2). Although a trichocyst has its own specific position in the surface pattern of the pellicle, mature trichocysts are sometimes but rarely found in the interior cytoplasm of the cell. If a fixative of slow penetration is employed, these trichocysts may also be discharged.

It is natural to ask what properties or advantages are conferred upon *Paramecium* as a result of its having the structures just described. If the ordered pattern of alveoli and corpuscular units in the pellicle system are

26

considered from this point of view, it is to be expected that the structure will give the pellicle rigidity without preventing flexure in certain directions. In the same way a papier-mâché egg-tray possesses rigidity because of the indentations which are impressed in it to hold the eggs separately. *Paramecium* may maintain its shape as a suit of chain mail is said to keep shape. It is also to be expected that the surface of *Paramecium* would be most easily capable of reversible deformation by folding either along the rows of cilia or between and parallel to these rows. Thus it should be easier to deform a paramecium by squashing it laterally than by bending its longitudinal axis. Live paramecia have been observed to deform themselves as they swim through a narrow gap. Pitelka (1965) has also suggested that an important function of the alveolar mosaic may be to protect the underlying cytoplasm from stresses due to chemicals which may permeate the cell membrane from the surrounding medium.

In *Paramecium* the cilia may beat forwards or backwards and when the animal reverses direction, as in the typical avoiding reaction, the cilia stop and the new direction of ciliary beating spreads from the end which will lead in the new movement (Parducz, 1959, Pitelka and Child, 1964). In the opinion of Bullock and Horridge (1965), however, these synchronous changes require the existence of no special conducting elements. The cilia of a freely swimming paramecium do not all beat in phase with each other, but rows of cilia in the same part of their beating cycle constitute the crests or troughs of metachronal waves; these have been observed to move over the cell surface in patterns which resemble the waves which can be seen in a field of wheat when the wind blows over it. The metachronal patterns are different for forward and backward movement in *Paramecium*. Bullock and Horridge (1965) admit that the coordination of the metachronal wave is difficult to understand but, having reviewed the available evidence, they assert that there is no need to invoke the existence of specialised structures which conduct excitation between the cilia. The system of kinetodesmal fibres, microtubules and fine fibrils below the surface of the pellicle may indeed conduct neuroid impulses but although a few microdissection studies have been performed, there is as yet no unequivocal demonstration that these structures do conduct excitation.

The structure and position of trichocysts and the nature of their discharge strongly suggest that their function is to deter a potential aggressor. They suggest an analogy with a harpoon gun used by whalers. Unfortunately for this line of thought there is no evidence that the discharge of trichocysts does actually deter an aggressor. The arch-enemy of *Paramecium* is *Didinium*, another protozoon which devours paramecia by engulfing them, but in this case trichocysts are ineffective. The suggestion has been made that a paramecium uses its discharged trichocysts to anchor itself to a surface while it feeds; but paramecia have been observed to swim about and feed at the same time. There seems, therefore, to be no satisfactory explanation for the function of trichocysts.

5 The gullet

In *Paramecium* the opening of the gullet or buccal cavity is funnel-shaped and tapers as it bends posteriorly into the interior of the animal to end at the cytostome or true mouth, below which food vacuoles are formed (Diagram 6). The general morphology of the gullet has been studied with the light microscope by Gelei (1934), Lund (1933, 1941) and Yusa (1957). It is about 30 μ long and 6 μ across. At its anterior and wider end, the gullet meets the vestibular region of the surface of *Paramecium* along the oval edge of the buccal overture. The vestibular region (Plate 15, fig. 1; Plate 16, fig. 1) is flared like a trumpet and lies at the posterior end of the oral groove (Plate 3, fig. 1). The vestibular region may be regarded as part of the pellicle and has a similar ultrastructure. The kineties of the vestibulum, however, are shorter and more strongly curved (Plate 2, fig. 1) and trichocysts are not found there. On the other hand the structures of the gullet or buccal apparatus are peculiar to this region. The infraciliature of the gullet is quite different from that below the pellicle. Although both regions have cilia, their ordered patterns of arrangement are different. The gullet is functionally distinct and has its own pattern of replication at fission. Stevenson (unpublished) has also shown that it is possible by isolation procedures to remove the gullet whole, without there being adherent fragments of pellicle (Plate 15, fig. 4). Moreover Ehret and Powers (1959) state that the gullet may be dissected out merely by compression.

Down the right-hand edge of the buccal overture there runs a row of cilia within a shallow groove, about 0.3 μ deep, between two cytoplasmic ridges (Plate 16, fig. 1; Plate 17, fig. 1). This structure is called the endoral membrane. According to Yusa (1964) there are twenty-two basal granules in the endoral membrane of *P. aurelia*. There are different opinions,

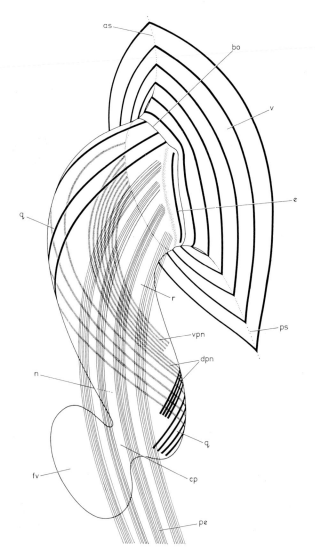

DIAGRAM 6 This three-dimensional diagram of the gullet of *P. aurelia* is viewed from a similar angle to the gullet in Plate 1. At its anterior end, the gullet meets the flared vestibulum (v) along the periphery of the buccal overture (bo). There is an anterior suture (as) and a posterior suture (ps) and a row of cilia is associated with the endoral membrane (e) to the right of the gullet. A row of small dots indicates the position of the small ridge (Plate 17, fig. 1) where the endoral membrane borders the ribbed wall (r). The close-packed eight rows of cilia which form the peniculus and the four more widely spaced rows of cilia which make up the quadrulus (q) each spiral down the length of

however, as to whether the endoral membrane has a single or double row of cilia. The electron micrographs here show that the basal granules are arranged in pairs (Plate 17, fig. 1) but only one basal granule of each pair appears to have a cilium above it.

Long rows of closely spaced cilia run in nearly parallel ribbons down the left side of the gullet. These rows show a dextral spiralisation as they run posteriorly down the buccal walls and a half turn is completed before the rows end near the bottom of the cavity (Diagram 6). All the rows of cilia begin at the buccal overture. On the anterior surface of the cavity and somewhat to the animal's left side there are four rows of cilia which constitute the ventral peniculus (Plate 17, fig. 1). Proceeding anticlockwise round the buccal overture from the point of view of an observer who is imagined to look into the buccal cavity, there are four more rows of cilia which constitute the dorsal peniculus (Plate 17, fig. 1). The ventral and dorsal peniculus together constitute a close-packed ribbon of cilia eight abreast (Plate 17, fig. 1; Plate 18, fig. 1) which has been shown to be 85 to 90 cilia long in the case of *P. bursaria* by Ehret and Powers (1959). The dorsal peniculus terminates at its posteriormost end rather closer to the bottom of the gullet than does the ventral peniculus.

Anticlockwise again from the peniculus, there are four rows of cilia which, for most of their length, are wider apart than are the rows of cilia in the peniculus (Plate 16, fig. 1; Plate 17, fig. 1). These four rows constitute the quadrulus. In a transverse section through *Paramecium* at about the middle of the buccal opening, the quadrulus is seen on the dorsal wall of the gullet (Plate 15, fig. 2; Plate 17, fig. 1). In this region the separation between adjacent rows of the quadrulus is at its widest. The rows are slightly closer

---

the gullet almost to its posterior end. The four rows of the dorsal peniculus (dpn) are longer than the four rows of the ventral peniculus (vpn). These rows of cilia are indicated by thick bands (black on the near side, hatched on the far side). The ribbed wall (r), which has no cilia or basal granules, is shown in the diagram on the nearer surface of the gullet while the naked ribbed wall (n) is on the far side. A food vacuole (fv) is shown during its formation and is connected to the posterior dorsal surface of the gullet by a short cytopharyngeal tube (cp). The postesophageal fibres (pe) run in the cytoplasm near the ribbed wall (r) down the length of the gullet and posteriorly beyond the bottom of the gullet.

at the anterior end but at the posterior extremity the rows are as closely packed as in the peniculus. Although for most of its length the quadrulus runs down the posterior and left-hand wall of the gullet, near the posterior end it veers across the ventral surface of the gullet to terminate on the right wall, at the end of the gullet and adjacent to the cytostome (Diagram 6).

Proceeding anticlockwise once more from the quadrulus there is an area of cell surface, without cilia or basal granules but with alveoli (Plate 16, fig. 1; Plate 19, fig. 1) which overlie a cytoplasmic surface which shows a series of ridges (Plate 17, fig. 1). This region resembles an area of pellicle without cilia or trichocysts and was termed 'the ribbed wall' by Ehret and Powers (1959).

Between the quadrulus and the ribbed wall, at a point somewhat more than half-way down the dorsal mid-line of the gullet, lies the apex of a triangular area of yet another type of surface organisation in the buccal wall, which extends down the posterior, dorsal region of the buccal wall to its base at the posterior end of the buccal cavity. This zone has cytoplasmic ridges but it is the only kind of cell surface without superficial alveoli (Plates 20 and 21). It can be called the 'naked ribbed wall' for here the cytoplasm is only bounded by a single unit membrane.

At the posterior dorsal extremity of the gullet, the naked ribbed wall is able to open along its lower dorsal mid-line to form the cytostome or the mouth, below which a short cytopharyngeal tube leads into a food vacuole (Plate 22). The posterior cilia of the quadrulus are placed so that they can assist bacteria to enter into the cytostome. When the food vacuole is full of bacteria, the cytostome closes. The cytopharynx and food vacuole are also bounded by a single unit membrane.

The peniculus and quadrulus together make up twelve rows of cilia (Plate 15, fig. 3) arranged in three groups of four rows each. Viewed from inside the buccal cavity there is a row of parasomal sacs immediately to the left of each fourth row of cilia (Plate 18, fig. 1), each parasomal sac being placed just anterior and to the left of the basal granule. Also, the basal granules of each fourth row of cilia are attached to root-fibrils (Schneider, 1964b), which the other rows lack (Plate 18). The root-fibrils extend about

0.5 μ to the same side of the basal granules as the parasomal sacs and slightly downwards to a lower level of cytoplasm; thus the root-fibrils of the peniculus point towards the quadrulus and those of the quadrulus are directed towards the ribbed wall. At high magnification it can be seen that each root-fibril is composed of a ribbon of about twelve short microtubules (Plate 18, fig. 3), which touch each other and which are arranged transversely, like a diagonally striped tie. There are alveoli between the basal granules of the peniculus and quadrulus (Plate 16, fig. 1). The dense layer immediately below the lower alveolar membrane is seen in tangential or oblique sections to consist of a hexagonal network of fine fibrils round the bases of hexagonally packed cilia (Plate 18, fig. 1). A few connecting fibrils run between this superficial layer of fine fibrils down to the proximal edge of each basal granule.

There are also connections between the proximal surface of the basal granules and a dense network of fibrils which runs in directions which are mostly parallel to the surface but which are below the level of the basal granules at depths down to about 0.8 μ beneath the surface. The fibrillar network shows a complicated but regular pattern when it is viewed in sections cut parallel to the surface as in the excellent micrographs of Schneider (1964b) and there are knot-like globular thickenings of a variety of sizes where diagonal fibrils intersect (Diagram 7). A similar fibrillar network was observed by Rouiller and Fauré-Fremiet (1957) in another ciliate, *Campanella umbellaria*.

Below the alveoli of the ribbed wall there are no basal granules. The nearest basal granules are those of the endoral membrane which forms a boundary between the ribbed wall and the vestibulum and there are also of course the basal granules of the quadrulus and peniculus which have a border with the ribbed wall. Although there are no basal granules beneath the ribbed wall, there is a knotted fibrillar network below it (Plate 18 and Diagram 7) which is similar to that which runs beneath the basal granules of the peniculus and quadrulus. This fibrillar network is clearest beneath that part of the ribbed wall which is near the endoral membrane. Beneath the opposite edge of the ribbed wall, which is near the quadrulus, there are more microtubules and less fine fibrils. Between the quadrulus and the ribbed wall and towards the anterior end of the buccal cavity, pouch-like folds in the inner alveolar membrane have been observed (Plate 16).

Below the naked ribbed wall (Plate 20) sheets of parallel microtubules run immediately beneath the surface and parallel to the length of the ridges. In transverse section (Plate 20, fig. 1) the microtubules appear as a row of small circles, one beneath the other and extending downwards from the apex of each ridge. The network of fine fibrils beneath the posterior region of the

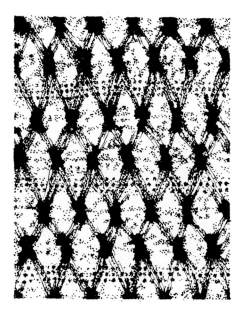

DIAGRAM 7 A drawing of the regular fine fibrillar network observed by electron microscopy beneath the surface of the ribbed wall of the gullet. The drawing is based on Plate 17, fig. 3, and on the fig. 11 of Schneider (1964b). Triple rows of small dense granules occur at intervals of about 400 mμ after every second row of the larger dense granular aggregations. The granules occur at the intersection of fine fibrils.

quadrulus also extends beneath the surface of the naked ribbed wall and individual bundles of fibrils there make connections with the microtubules beneath the ridges (Plate 21, figs. 3 and 4). The sheets of microtubules (Plate 20, fig. 3) may run in straight lines beneath the surface for distances of at least 6 μ or they may veer away from the surface into the endoplasm in whorls (Plate 21, fig. 4).

Between the sheets of microtubules there are scattered numerous disc-shaped bodies, 0.3 μ in diameter and about 60 mμ thick (Plate 20 and Diagram 8) which are nearly always oriented so that the plane of the disc is parallel to the sheets of microtubules. These bodies are bounded by unit membranes. There is, however, a dense layer immediately within the unit

34

membranes (Plate 20, fig. 6) but the innermost part of the disc is less dense. The disc bodies are seldom found far from the cytopharyngeal region. The functional significance of these bodies is unknown but they may play a part in the engulfment of food.

About five hundred microtubules, closely packed in bundles of about twenty-five, run beneath the ribbed wall of the gullet and then continue into the cytoplasm posterior to the gullet (Plate 19). The bundles of microtubules are at least 35 μ in length. The microtubules are 25 mμ in diameter and closely packed in the bundles with a centre to centre distance of about 35 to 40 mμ (Plate 19, fig. 3). Each microtubule runs parallel to the length

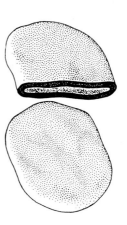

DIAGRAM 8   A diagram of the disc-shaped bodies which are found in the cytoplasm below the naked ribbed wall of the gullet. The disc-shaped body to the top of the figure is shown with half cut away to show how the elongated profiles arise in thin sectioned material (see Plate 20, figs. 1, 4 and 6). The disc-shaped bodies are 0.3 μ in diameter and about 60 mμ thick and have a dense layer immediately within their limiting membrane and a less dense central region.

of the bundle (Plate 19, fig. 2). Pitelka (1967) observed these microtubular bundles in *P. multimicronucleatum* and called them post-oral fibres. They correspond to the postesophageal fibres of Lund (1933) which were observed by light microscopy; Lund (1941) suggested that the fibres are involved in a process by which food vacuoles, once they are formed, are passed rapidly in a posterior direction into the cytoplasm.

The anterior origin of the postesophageal fibres has not been determined with certainty. Electron micrographs of a newly developed gullet have been obtained which show tightly packed bundles of microtubules originating from a row of basal granules (Plate 39, figs. 2 and 4). In this case the microtubules may be the anteriormost end of the postesophageal fibres but the location of the basal granules is uncertain.

D

Also in the cytoplasm beneath the ribbed wall and beneath the naked ribbed wall at a depth of between 0.5–1 μ below the surface of the buccal wall, there run a few long fibrous cords about 0.25 μ thick. In well-fixed material no definite ultrastructure can be seen within the cords (Plate 20, fig. 1) but in poorly preserved material a mass of fine fibrils has been seen in a region corresponding to that of the cords (Plate 20, fig. 5; Plate 21, fig. 1).

There are no kinetodesmal fibrils beneath the surface of the buccal cavity. This statement can be made with some certainty because of the absence of the characteristic cross-banding by which kinetodesmal fibrils may be recognised. The parasomal sacs always occupy the same position with respect to the nearest basal granule, both at the surface of the buccal cavity and in the pellicle (as described in Chapter 4). On the other hand the rest of the infraciliature in the two regions is different. The rootlet fibres, sheets of microtubules and the fibrillar networks of the buccal region do not correspond to the arrangement of microtubules and fine fibrils below the pellicle. Although it is possible to regard the pellicle and buccal cavity systems as derived from the same basic system of surface structure, the ultrastructural dissimilarities are still considerable.

6 Food vacuoles

In *Paramecium*, as in all ciliates, the food vacuoles (Plates 22 and 24) form an essential part of the feeding apparatus and have a function similar to the stomach and intestines of higher organisms. For this reason Volkonsky (1934) called them gastrioles and the name, which did not gain general acceptance, at least had the merit of avoiding confusion with the contractile vacuoles.

Paramecium has always been a favourite animal for the study of food vacuoles by light microscopy, as the reviews of Kalmus (1931) and Wichterman (1953) make clear. Food vacuoles in *Paramecium* have been examined under the electron microscope by Jurand (1961) and Schneider (1964b) and in certain other ciliates by Carasso, Favard and Goldfischer (1964) and Favard and Carasso (1964). One of the objects of research on food vacuoles has been to understand how proteolytic and other enzymes associated with digestion are transported from the ribosomes, where presumably they are synthesised, to the site of their action which is inside the food vacuole. It is also necessary to understand how nutrients are ingested from the food vacuole to the cytoplasm and how the materials which *Paramecium* cannot assimilate are eliminated.

Food vacuoles are formed beneath the cytostomal region of the buccal cavity when the sack-like cytopharyngeal tube is pinched off by a wall of cytoplasm which closes above it (Plate 22). At this stage the bacteria trapped within the vacuole appear intact and normal. Subsequently the age of a food vacuole may be roughly estimated by observing the state of dissolution of the bacteria within it. A very young food vacuole may be surrounded by a few of the cytostomal disc-shaped bodies in the cytoplasm but these show no association with the wall of the vacuole. On the other hand, a young food vacuole is often surrounded by numerous mitochondria, elements of endoplasmic reticulum and small cytoplasmic vesicles about 0.1–0.2 μ in diameter. Both the vesicles and the food vacuoles are bounded by a single unit membrane.

The sites of activity of the enzyme acid phosphatase are particularly easy to demonstrate cytochemically using the light microscope. The presence of this enzyme in a particular region of the cell may often indicate the presence of proteolytic enzymes or digestive hydrolases. Rosenbaum and Wittner (1962) and Müller (1962) found acid phosphatase and other hydrolases localised round the periphery of food vacuoles in *P. caudatum* and *P. multimicronucleatum*. They further demonstrated a correlation between the sites of acid phosphatase activity and the position of the cytoplasmic vesicles which are stainable *in vivo* with neutral red. The neutral red reaction round the food vacuoles of Protozoa has been studied since the early work of Provazek (1898) and more recently in *Paramecium* by Mast (1947).

The work of Carasso, Favard and Goldfischer (1964) with the electron microscope not only supported the findings of Rosenbaum and Wittner (1962), but showed that acid phosphatase in the peritrichous ciliate *Campanella* is found in the cisternae of the endoplasmic reticulum surrounding young vacuoles. The electron micrographs of Schneider (1964b) showed 'enzyme granules' as irregular vesicles 0.1–0.5 μ in diameter (similar to those of Plate 23, fig. 3) around the young food vacuoles of *Paramecium aurelia* but no clear relationship was found between the 'enzyme granules' and the endoplasmic reticulum. Schneider suggests that the contents of the 'enzyme granules' are discharged into the food vacuoles. Müller, Röhlich, Tóth and Törö (1963) found acid phosphatase activity localised in small vesicles attached to the young food vacuole of *P. multimicronucleatum* or confined to the membrane in vacuoles of a later stage.

In the older food vacuoles, bacteria may be observed in different stages of digestion. The bacterial cell membrane is often expanded and lifts away from the cytoplasm. Later the food vacuole is filled with a mass of colloidal and fibrous matter which is presumably derived from disintegrated bacteria (Plate 24, fig. 1). The food vacuoles at this stage appear to be somewhat larger than formerly. The unit membrane of the vacuole assumes a wavy outline with frequent sac-like protuberances which extend into the cyto-plasm. These protuberances are sometimes flask-shaped with comparatively narrow necks and wider ampoules about 0.2 μ in diameter (Plate 24, figs. 1 and 5). There are also numerous small secondary vacuoles, also about 0.2 μ in diameter, within the cytoplasm round the main vacuoles. Sometimes these secondary vacuoles appear flattened or assume the shape of a hollow ball which is deformed by pushing one side inwards until it approaches the opposite side (Plate 24, fig. 4). The old food vacuoles are not surrounded by concentrations of mitochondria or endoplasmic reticulum.

The infolding of the membrane of the older food vacuole to form small secondary vacuoles is almost certainly an example of a process called 'pino-cytosis' (Lewis, 1931), which allows small quantities of the vacuolar fluid containing food material to be transferred into the cytoplasm where the process of food absorption may be completed. Pinocytosis has been exten-sively studied in *Amoeba* where the cell membrane is involved (see Holter, 1959, 1963), and has been observed in the food vacuoles of *Paramecium aurelia* by Jurand (1961) and Schneider (1964a). By observing the electron-opaque substances taken into the food vacuoles, Schneider (1964a) in *Paramecium* and Favard and Carasso (1963) in the peritrichous ciliate *Epistylis anastatica* have confirmed that the process of pinocytosis does remove solid material from the vacuole into the surrounding small vesicles. Even without the use of electron-dense tracer substances, however, solid material may be seen inside the cytoplasmic vesicles formed by pinocytosis (Plate 24, fig. 3). Favard and Carasso (1964) also present evidence that undigested solid material is returned from the flattened small vesicles back into the food vacuole.

The oldest food vacuoles (Plate 24, fig. 6) are not as rounded as those at the earlier stages and some have an irregular outline characteristic of a partially collapsed state. They contain few or no undigested bacteria but contain some quantity of fine granular matter together with bacterial 'ghosts' which are presumably bacterial cell walls. Such vacuoles come to a position immediately under a specially differentiated excretory region of the cell surface called the cytopyge.

The cytopyge (Plate 1) occupies a fixed position in the pellicle (Plate 2, fig. 3) on the mid-ventral suture line about 30 μ posterior to the opening of the gullet. Von Gelei (1939) published an excellent photograph of the cytopyge region of the pellicle using light microscopy and a silver-stained specimen. The cytopyge has been observed in section under the electron microscope by Schneider (1964a), both when an old food vacuole was positioned immediately beneath it and when the nearest vacuole was some distance away. Under the electron microscope the cytopyge appears as a small area of cell surface, about 1 µ across, which is only bounded by a cell

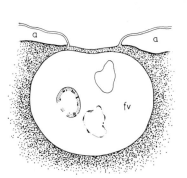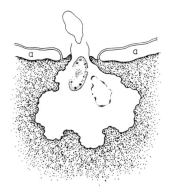

DIAGRAM 9   These two diagrams which are based on the electron micrographs of Jurand (1961) and Schneider (1964b) show the expulsion of an old food vacuole through the cytopyge, which is a small opening in the pellicle on the mid-ventral suture line (Plate 2, fig. 3) where the cytoplasm is only bounded by a plasma membrane. Alveoli (a) occur on either side of the cytopyge. In the diagram to the left, the old food vacuole (fv) is positioned below the cytopyge. In the diagram to the right, the thin layer of cytoplasm above the food vacuole has been ruptured to allow it to be discharged but the plasma membrane is intact on each side of the vacuole.

membrane but which is surrounded by the alveoli of the corpuscular system of the pellicle. Jurand (1961) produced an electron micrograph which shows a food vacuole with its contents being expelled. It is obvious that immediately before the contents of a food vacuole are released into the surrounding water through the cytopyge (Diagram 9) the thin layer of cytoplasm above the food vacuole must be momentarily breached; but because the cytoplasmic bridge at this stage is so thin (less than 0.1 µ) little cytoplasm is likely to be lost before the cell membrane reforms in a position which allows the food vacuole to fit into the cytopyge opening and later to release its contents outside the cell.

40

7 Contractile vacuoles

In *Paramecium aurelia* there are two contractile vacuoles (Diagram 10) which occupy fixed positions under the surface on the dorsal mid-line, one near the anterior and the other near the posterior end of the cell (Plate 1). The discharge channel of the contractile vacuole (Plate 25) appears to replace one corpuscular unit of the pellicle (Plate 11, fig. 6). Contractile vacuoles periodically increase in volume at diastole; they discharge their contents at systole and then begin to increase in volume again. The pulsations of the contractile vacuole are easy to observe in polar view by light microscopy if the animals are first immobilised by just sufficient pressure between a slide and coverslip. In normal pond water a contractile vacuole may discharge once every 6–10 sec. Diastole is normally of 5–10 sec duration and therefore lasts longer than systole which may take about 2 sec. There is a steady rate of expansion of the contractile vacuole in diastole but its collapse is sudden and occupies only about 0.2 sec. The contractile vacuole cannot normally be observed by light microscopy in systole because as the vacuole collapses it becomes too flattened to be visible and only the fully dilated nephridial canals can be seen extending outwards from the position of the vacuole.

Almost certainly the vacuole performs an osmoregulatory function. Water tends to accumulate in the cytoplasm by osmosis across the cell membrane and by ingestion from food vacuoles; it may also be formed as a metabolite in the cytoplasm.

The pulsation period of contractile vacuoles is known to depend on the temperature and osmotic pressure of the medium. With *P. caudatum*, for instance, Herfs (1922) showed that as the temperature was increased from 14° C to 23° C the pulsation period decreased between two and three times,

41

whereas when the concentration of sodium chloride added to pond water was increased to 1 per cent the pulsation period increased from 6 sec to 3 min. These observations support the conclusion that contractile vacuoles are osmoregulatory organelles. Additional support is provided by consideration of the marine protozoa, which either do not possess such a vacuole or, in cases where it is present, the vacuole has a considerably longer pulsation period. Howland (1924) and Weatherby (1927) considered that contractile vacuoles also excrete nitrogen metabolites such as uric acid and urea; this, however, is difficult to confirm or disprove as it is not easy to analyse the contents of a contractile vacuole.

In *Paramecium* the contractile vacuole (Plate 25, figs. 1 and 4) is about 8 μ in diameter, when fully expanded. Seven to ten nephridial canals (Diagram 10 and Plate 27) extend radially from the main vacuole for a distance of about 12 μ into the surrounding cytoplasm. The nephridial canals connect with the main vacuole through a short narrow injector canal which is about 0.5 μ long and 0.1 μ in diameter. This widens out into the middle section of the nephridial canal, called the ampulla, which is about 4 μ in length. The distal part of the nephridial canal is longer and narrower than the ampulla. The ampullae are most easily observed by light microscopy at systole but in fact they can be seen 1 sec before the collapse of the contractile vacuole and also before they shrink to invisibility, that is for 2 sec after the contractile vacuole begins its enlargement.

After the main vacuole has discharged at systole, the main distal part of the nephridial canal opens to its widest diameter of about 0.4 μ (Plate 27, fig. 4) and can be seen in the light microscope as a dark line which leads into the ampulla. The canal is surrounded by a dense sponge-like network of minute nephridial tubules (Schneider, 1959b) which connect with the main canal and branch off in all directions into the cytoplasm. When the main vacuole is fully expanded at diastole, the nephridial canals are at their narrowest (Plate 27, fig. 1) and no connection can be observed between the minute nephridial tubules and the nephridial canal (Schneider, 1960a). Thus the main vacuole is filled by water which appears to be forced by the peristaltic movements of the nephridial canals to flow from the minute cytoplasmic tubules to the distal section of the nephridial canal and from

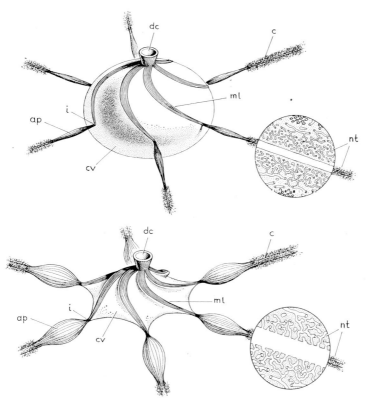

DIAGRAM 10   The diagrams are three-dimensional representations of a contractile vacuole in diastole (above) and systole (below) and have been reconstructed from electron micrographs (Plates 25 to 27) and from Schneider (1960a). Six nephridial canals (c) are radially disposed round the main part of the vacuole (cv). The nephridial canals are connected to the main vacuole by a narrow injector canal (i) which opens into a wider ampulla (ap). Along most of its length the nephridial canal is covered by a network of narrow nephridial tubules (nt). An enlarged circular inset into a nephridial canal shows that when the contractile vacuole is in diastole (above) the canal is at its narrowest and the nephridial tubules (nt) do not connect with it. The enlarged inset in the lower diagram shows that when the main vacuole is in systole the nephridial canal is fully expanded and the nephridial tubules connect with it. Ribbons of microtubules (ml) run from the wall of the discharge channel (dc) down the wall of the contractile vacuole and then along the outside of the nephridial canals.

there into the ampullae and via the injector canals into the main chamber.

The main vacuole is bounded by a unit membrane. The roof surface of the vacuole is reinforced by ribbons of tubular fibrils, each 25 mμ thick (Plate 26, figs. 5 and 6) which probably make the structure contractile. At systole the ribbed roof of the vacuole is thrown into a number of folds (Diagram 10) and the floor of the vacuole is raised to be almost in contact with it (Plate 25, figs. 2, 5, and 6). The tubular fibrils also extend around the discharge channel (Plate 25, figs. 3 and 6; Plate 26, figs. 1, 2, and 4) which changes diameter with the pulsations of the vacuole. This channel is closed by an inverted thimble-like structure inserted into the pellicle (Plates 25 and 26). The septum itself is 40 mμ thick, and appears to be formed of two membranes (Plate 26, fig. 3), the outermost one being continuous with the pellicle and the innermost being a unit membrane continuous with the membrane of the vacuole.

It seems obvious that the septum serves as a valve which allows the content of the contractile vacuole to be discharged at systole but prevents water entering from outside during diastole. All the micrographs, however, whether from fixations made at diastole or fixations made at systole, have so far shown the septum intact. A possible explanation is that the septum becomes ruptured at systole but is then immediately repaired.

8 The cytoplasm and its organelles

The interior cytoplasm or endoplasm of *Paramecium* has been studied less than the superficial cytoplasm or ectoplasm, because the latter contains organelles and structural modifications peculiar to *Paramecium*, while the endoplasm contains organelles essential to cells in general. Organelles such as mitochondria and the rough endoplasmic reticulum are recognised as essential to the metabolic activity of nearly all eukaryotic cells and there is usually no reason to study them in any particular species.

The endoplasm of *Paramecium* contains mitochondria, endoplasmic reticulum, ribosomes, glycogen granules and lipid drops. It also contains rounded vesicles which are approximately 0.2 μ in diameter and some of which contain fine irregular fibrous material. There are a few microtubules which do not appear to be associated with any of the structures of the pellicle or gullet. There are also some tubules of 50 mμ diameter. It is difficult to observe any other elements of cytoplasm but there appear to be irregular fibrous elements near to the limits of resolution.

The mitochondria of *Paramecium* (Plate 28, fig. 1) are typical of those found in the majority of Protozoa. They assume a rather irregularly rounded rod shape and are about 2 μ long and 0.7 μ in diameter. They are bounded by two membranes separated by a less dense layer of about 8 mμ (Plate 29, fig. 4). Internally there are numerous cristae in the form of cylindrical tubules which may be straight or curved (Powers, Ehret and Roth, 1955; Powers, Ehret, Roth and Minick, 1956; Vivier, 1966). Each tubule is about 50 mμ in diameter and is bounded by a membrane which at one end of the tubule is continuous with the inner limiting membrane of the mitochondrion. Each of the two limiting membranes of the mitochondrion usually appears as a dark boundary about 6 mμ thick (Plate 29, fig. 4). Schneider (1959a) showed, however, that the membranes forming the walls of the cristae in mito-chondria of *Paramecium* could be resolved into two dense layers enclosing a less dense one. In our electron micrographs these membranes sometimes show a beaded appearance produced by corpuscular units about 8 mμ in diameter. The ultrastructure of these membranes may therefore be similar to that of the membranes in the cristae of metazoan mitochondria as studied by Sjöstrand (1963).

Between the tubules of the mitochondria, the matrix in *Paramecium* is dense and granular. In some sections there are to be found within the matrix a few patches of even denser fibrillar material (Plate 28, fig. 1) whose chemical nature is unknown. It is interesting to note that double-stranded DNA has been isolated from mitochondrial fractions of *P. aurelia* by Suyama and Preer (1965). Schneider (1960a and b) and Wohlfarth-Bottermann and Schneider (1961) studied the degeneration of mitochondria caused by doses of X-rays given to *Paramecium*.

Grimstone (1961) has pointed out that it is incorrect to suppose that the presence of tubular cristae is a feature by which protozoan mitochondria may be distinguished from those of other groups; the flat, shelf-like cristae occur in the mitochondria of several flagellates (Sager and Palade, 1957; Manton, 1959; Roth, 1959) as well in the heterotrichous ciliate *Spirostomum ambiguum* (Randall, 1957), while tubular cristae are found in certain other types of cell (Sedar and Rudzinska, 1956; Rouiller, 1960; Novikoff, 1961) including those of certain metazoa (Sedar and Rudzinska, 1956).

The mode of formation of mitochondria is as controversial a subject for protozoa as it is for the metazoa. Mitochondria of *Paramecium* may be derived from the division of existing mitochondria (Wohlfahrth-Botter-mann, 1956a, 1956b, and 1957), but there is also evidence that the mito-chondria may develop from small undifferentiated membrane-limited vesicles which originate *de novo* in the endoplasm (Wohlfahrth-Botterman, 1956a, 1956b, 1957, 1958b). It was proposed that mitochondria of *Para-mecium* are discharged into the cytoplasm from the macronucleus (Ehret and Powers, 1955) but this idea has received no support.

Mitochondria with modifications from the usual structures are quite easy to find in *P. aurelia* (Plate 40). The outer membrane of a mitochondrion is

46

sometimes found extended into the cytoplasm, enclosing fine irregular fibrogranular material and with no cristae within the bud-like extension (Plate 40, figs. 1 and 2). Similarly modified mitochondria were observed by Ehret and de Haller (1963) in *P. bursaria*. For *P. aurelia*, a series of electron micrographs can illustrate the development of new mitochondria from their origin as the bud-like extensions which break off from parent mitochondria to form vesicles (Plate 40). Tubular cristae are seen at various stages of development within the vesicular bodies (Plate 40, figs. 3–6). There would appear to be at least as much evidence to support this hypothesis for the reproduction of mitochondria as there is for the alternatives which have been proposed.

In electron-microscope studies of cells other than *Paramecium* almost every organelle has been implicated as the source or site of development of mitochondria but none of the evidence provides more than a plausible hypothesis. Since it is the function of mitochondria to make energy in the form of ATP available within the cell, it is not at all surprising to find mitochondria crowded round organelles which have an energy requirement. In *Paramecium* high concentrations of mitochondria are found immediately beneath the fibrillar bundles under the pellicle (Plates 5 and 6) where they are near both the cell surface and the basal bodies of cilia. Whatever the mechanism for the beating of cilia, energy will be required for the constant motility observed. Oxygen must also be more readily available near the cell surface and energy transfer to the cilia will therefore operate at maximum efficiency when the mitochondria are near the surface.

In the cytoplasm of *P. aurelia*, there is an abundance of food storage granules which are almost certainly glycogen. After favourable conditions of preservation, the granules have a diameter of about 40 mμ and a somewhat irregular outline (Plate 29, figs. 2–4). They are therefore larger than the ribosomes (Plate 29, fig. 2) which are only about 15 mμ in diameter. Perry and Sinden (personal communication) have demonstrated that in *P. aurelia* these granules show a substructure consisting of particles about 4 mμ in diameter when they are treated with periodic acid before staining with lead citrate (Plate 29, fig. 5). In this respect, therefore, they behave identically to similar granules in amphibian embryos which Perry (1967) has identified as glycogen.

47

In the cytoplasm of *P. caudatum* lipid droplets have been observed with the electron microscope by Schneider (1959a, 1960a) and André and Vivier (1962) as irregular granules about 0.5 μ in diameter, stained deeply by osmium tetroxide. In *P. aurelia* refractile lipid droplets, known to be of low specific gravity appear in the light micrographs of Preer (Plate 51, figs. 1 and 2). Our electron micrographs, however, show no lipid in the form of dark granules, although homogenous material bounded by a membrane has been observed as drops of similar volume to a mitochondrion (Plate 29, fig. 1). These may be lipid but they do not darken by osmic fixation. Casley-Smith (1967) reports that free lipids are readily dissolved by the embedding media normally employed for electron microscopy.

In *P. aurelia* the rough endoplasmic reticulum takes the form of individual flattened vesicles with ribosomes attached to them (Plate 29, figs. 1 and 2). The endoplasmic reticulum is never found organised into well-ordered stacks of flattened vesicles, all parallel to each other, as in the cells of certain glands in vertebrates. The individual vesicles in *Paramecium* are mainly concentrated in the cytoplasm together with the mitochondria near the cell surface (Plate 6 and Plate 40, fig. 6).

It is probable that there are many ribosomes which occur freely scattered throughout the cytoplasm of *Paramecium* and which are not bound to membranes. Sommerville and Sinden (1968) estimated that free ribosomes are approximately 70 per cent of the total number of ribosomes. In many electron micrographs, however, it is difficult to identify free ribosomes.

Jurand, Gibson and Beale (1961) showed that 80 per cent of *P. aurelia* were able to survive a treatment for 12 h in exhausted culture medium which contained 0.5 mg per ml of ribonuclease. The treated individuals lost most of the pyronin-staining material which normally is found throughout the cytoplasm; they also lacked virtually all the granules which have dimensions similar to ribosomes, although the membranes were still present. Glycogen was probably not preserved by the preparative techniques employed but nevertheless the work provided evidence in favour of the presence of numerous free ribosomes in the cytoplasm of the normal *Paramecium* cell.

Smooth endoplasmic reticulum in the form of flattened vesicles is rarely observed in *Paramecium*, but smooth rounded vesicles are common. Grim-

48

stone (1961) states that most ciliates have no Golgi material, and it is true that *Paramecium* appears to lack any structure resembling a well-developed dictyosome or a Golgi body. A few profiles have been seen (Plate 28, fig. 3) which suggest a poorly developed Golgi complex. If Golgi material in *Paramecium* were to consist of small aggregates of smooth and rounded vesicles in the endoplasm, then it would be difficult to recognise it on purely morphological evidence. The endoplasm, however, does contain some irregular aggregates of vesicles, each between 0.1 and 0.5 µ in diameter and bounded by a smooth-walled membrane (Plate 28, fig. 2). These groups of vesicles are similar to the vesicular type of Golgi complex observed in some insect cells by Jacob and Jurand (1965) and Jurand, Simões and Pavan (1967). On the other hand such vesicles might represent the sites of extracted lipid.

In some parts of the cytoplasm there are stacks of membranous tubules each 50 mµ in diameter and about 1.3 µ long. They were called the ortho-tubular system by Ehret and de Haller (1963). In a typical stack (Plate 28, fig. 4) about twenty such tubules are found hexagonally packed with a centre-to-centre distance of 100 mµ. Outside each tubule, knob-like projections of about 10 mµ diameter appear to be joined to the membrane by a short neck about 5 mµ long, suggesting that these tubules may have sub-units of a similar size and shape to those observed by Fernández-Morán, Oda, Blair and Green (1964) in mitochondrial membranes. The stacks of tubules are not distributed uniformly in the cytoplasm but are found near the nephridial canals where they were observed in the furrow region at fission (Plate 35, fig. 5). No continuity is known to exist between nephridial canals and these tubules.

In some cultures of *P. aurelia* abundant irregular crystals are found scattered throughout the interior cytoplasm. They have been observed by light micro-scopy (Kalmus, 1931 and Plate 30, figs. 1 and 2). They are also found in electron micrographs (Schneider, 1963 and Plate 30, fig. 3). They may be at least 5 µ across or they may be much smaller. The crystals are enclosed in individual vacuoles separated from the cytoplasm by a membrane. They are of unknown nature and there is no evidence that they are food storage granules or that they represent a waste product.

9 The nuclei

Two different kinds of nuclei which differ in size and in function are found in *Paramecium* (Plates 1 and 31) and in other ciliates. In each animal there may be one or more micronuclei and there is usually one macronucleus. The macronuclei are concerned with the control of growth and morphogenesis (Nanney and Rudzinska, 1960). The macronucleus of *P. aurelia* has been shown by Sonneborn (1946) to contain the genes which determine the phenotype of the organism. No growth occurs in the absence of a macronucleus (Balamuth, 1940; Weisz, 1951), but some ciliates do retain the capacity for growth and morphogenesis in the absence of a micronucleus.

    P. aurelia cannot undergo more than one or two fission cycles without the macronucleus, but without a micronucleus the animals may survive for over a hundred fissions and are even capable of mating (Sonneborn, 1938a). The micronuclei periodically differentiate into macronuclei in the course of the nuclear reorganisation that follows conjugation or autogamy. Conjugation and autogamy result in rejuvenation of clones derived from a single individual. On the other hand macronuclei are incapable of giving rise to micronuclei. The biological rôle of the micronucleus is to permit the genetic recombination which occurs at conjugation and at the first autogamy following conjugation.

    In *P. aurelia* there are two micronuclei which before division are about 3 μ in diameter (Plate 31, fig. 2) and which lie close to the macronucleus (Plate 31, figs. 4 and 5). Light microscopy of fixed and stained material shows the micronuclei as hollow achromatic spheres bounded by a thin membrane and containing a central chromatic core (Sonneborn, 1947). Under the electron microscope (Plate 31, fig. 1) they have a similar appearance, except that the dense core often appears as a hollow or cup-shaped object immediately surrounded by a less dense granular zone, while the bulk of the micronucleus is filled with fibrous material of low electron density. The nuclear membrane is double and almost certainly contains pores (Plate 31, fig. 3). In a few electron micrographs the outer layer of the nuclear membrane appears to be continuous with sac-like membraneous protrusions extending into the cytoplasm. These observations are somewhat similar to those which demonstrate the origin of endoplasmic reticulum in metazoa (Waddington and Perry, 1962; Jurand, 1962).

The macronucleus is large and ovoid. In *P. aurelia* it normally is about 35 μ in length and 12 μ transversely across its diameter. Its volume is therefore roughly 180 times that of a micronucleus in interphase. By microphotometry the macronucleus of *P. aurelia* has been estimated to contain 430 times as much DNA as the micronucleus (Woodward, Gelber and Swift, 1961; Woodward, Woodward, Gelber and Swift, 1966). The macronucleus is considered to be endopolyploid. In support of this view is the observation of Sonneborn (1957) that the macronucleus can regenerate from fragments formed by disintegration of the old macronucleus at conjugation, provided that the development of new macronuclei is suppressed. Sonneborn (1956) has also shown that heterozygous macronuclei retain their heterozygosity after hundreds of cell divisions. This latter observation may either be taken as evidence that in macronuclear divisions the daughter chromosomes are regularly distributed to the daughter nuclei as in a normal mitosis, or that the macronucleus is made up of diploid subunits as Sonneborn proposed. Diploid subunits within a macronucleus of *Paramecium* have not been observed under the electron microscope but many individual spindles were noted within the dividing macronucleus of another ciliate *Campanella umbellaria* (Carosso and Favard, 1965). It is also possible that the macronucleus is polyploid with respect to an incomplete chromosomal set in which only somatic determinants are included.

Electron microscopy has shown that in the macronucleus of *Paramecium aurelia* there are randomly scattered, dense fibrigranular aggregates which are usually between 0.5 μ and 1.0 μ in diameter (Plate 32 and 33) and which were termed 'large bodies' by Jurand, Beale and Young (1962). Sonneborn (1953) estimated that there were about a thousand of these large bodies in a fully developed macronucleus. They commonly show a hollow centre of lower density which may be between 0.2 and 0.5 μ in diameter (Plate 33) while the granules in the large bodies are about 15 to 20 mμ in diameter. For every one of these large bodies there may be about a hundred smaller dense aggregates (Plates 32 and 33) of somewhat similar electron density but with a more fibrillar and less granular ultrastructure; these have been termed 'small bodies' and are only between 0.1 and 0.2 μ in diameter (Plate 33, fig. 2). Outside the large and small bodies is a background matrix of low electron density fibrillar material with fibril diameters of about 5 mμ and less.

Nanney and Rudzinska (1960) called the large bodies nucleoli and the small bodies 'chromatin bodies'. At present there are several pieces of evidence in favour of the view that the large bodies contain RNA, and rather less evidence to support the view that the 'small bodies' (as distinct from the fibrigranular background) represent the chromosomes. The appearance of the dense masses varies with the state of the macronucleus; for instance in an old paramecium culture, in which the animals were not allowed to undergo autogamy for 45 days, large dense granular aggregates or 'large bodies' of up to 3 μ in diameter, were observed (Jurand, Beale and Young, 1962). Treatment with ultra-violet light produced similar aggregates (Kimball, 1949). In very young macronuclear anlagen of *P. aurelia* (Plate 49, fig. 1) where presumably the rate of synthesis of DNA is high, there are no granular aggregates to be seen at all (Jurand, Beale and Young, 1964) and 12 h later, when the anlagen have increased in size and contain many large bodies, the small bodies are still not present (Plate 49, fig. 5). In the macronucleus of the ciliate *Euplotes* the large and small dense bodies are similar to those in *Paramecium* but in *Euplotes*, DNA synthesis occurs in reorganisation bands which approach the middle of the macronucleus from either end (Gall, 1959). In the reorganisation bands no dense aggregates can be observed (Fauré-Fremiet, Rouiller and Gauchery, 1957). It seems probable, therefore, that there are no dense aggregates associated with macronuclear chromatin when DNA is being synthesised.

Recently Wolfe (1967) spread macronuclei from *P. aurelia* on an air-water interface before examination by electron microscopy, and showed that the small bodies of 0.1 to 0.2 μ diameter contain DNA and are joined together by means of fibrils about 10 mμ in diameter. The latter may be paired to form coils of 25 mμ diameter within the small bodies.

Jurand, Beale and Young (1962) using light microscopy and sectioned material, observed in interfission macronuclei, pyronin-positive 'vacuoles' about 1 μ in diameter and containing centres which were positive to methyl-green or Feulgen staining and were near to the limit of optical resolution. The identification of the large granular masses ('large bodies' observed by electron microscopy) with the pyronin-positive vacuoles leads to the conclusion that the former contain RNA and that probably their less dense centres, observed by electron microscopy, contain DNA. This represents a re-interpretation of the results of Jurand, Beale and Young (1962) who also showed that, after treatment of a macronucleus with DNA-ase and subsequent observation by ultraviolet light, considerable extraction had taken place and that coarse granular masses about 0.6 μ in diameter were left behind. It is easy to assume that the coarse granular masses correspond to the large bodies which contain RNA. The results of the silver-Feulgen method, also used by these authors, do not, however, fit this reinterpretation and the method may be unreliable. Autoradiographic observations by electron microscopy after labelling the DNA with tritiated thymidine support the view that the large bodies do not contain a high concentration of DNA (Jurand and Jacob, unpublished). This result should be supported by chase experiments and by the use of labelled precursors of RNA.

A method for the isolation of large numbers of macronuclei *in vitro* has recently been developed by Stevenson (1967b) so that macronuclei may now be subjected to biochemical analysis. Examination of the isolate in the electron microscope by Jurand (Plate 32, fig. 2) showed that the macronuclei were reasonably free from contamination by other cytoplasmic constituents and that they retained their general morphology although their nuclear membranes were lost.

The macronucleus has a conventional double nuclear membrane with pores (Plate 32, figs. 3 and 4).

10 Binary fission and the fission cycle

Paramecia can increase their numbers only by binary fission, a process that was described for *Paramecium aurelia* by Hertwig (1889) and by Sonneborn (1947). A transverse constriction furrow grows inwards at the animal's equator (Plate 39, fig. 1; Plate 41, fig. 6; Plates 36 and 37) to divide it into two daughter animals, each half the volume of the parent, the volume being made good by growth before the next fission. Immediately after fission, the daughter animals are shorter and more rounded than their parent and they also lack an oral groove. The two daughter paramecia, however, are not identical to each other. One daughter cell, sometimes called the 'proter', is derived from the anterior half of the parent cell and undergoes subsequent growth and reorganisation to provide itself with a posterior end. The other daughter cell, sometimes called the 'opisthe', is derived from the posterior half of the parent and must regenerate an anterior end. Of the two mature contractile vacuoles, one passes to each daughter animal and one new contractile vacuole must develop in each case. New contractile vacuoles are first observed towards the end of the interfission period but before the constriction of the fission furrow (Hanson, 1962; Kaneda and Hanson, 1967). Their morphogenesis has not been studied.

The two micronuclei of *P. aurelia* undergo mitosis nearly synchronously during the last hour before fission. Electron microscopy of thin-sectioned material reveals new details of the mitosis. The dense granular region which occupied a central position in interphase is more dispersed in prophase (Plate 35, fig. 2). There is also a microtubular zone just inside the nuclear membrane at this stage and earlier (Plate 35, fig. 1). At the metaphase plate stage (Plate 34, fig. 1) small densely coiled chromosomes are attached by localised kinetochore regions to microtubular elements of the spindle (Plate

54

35, fig. 3). The metaphase plate is 6 μ in diameter and the individual chromosomes appear so small by light microscopy that their number has never been reliably estimated. Electron microscopy confirms that the nuclear membrane remains intact at metaphase and throughout the subsequent stages of mitosis (Plates 34 and 45). This result was recorded by the early light microscopists (Sonneborn, 1947) but because the nuclear membrane is too thin to be resolved by light microscopy it is difficult to understand how the observations could have been made. At anaphase and telophase the spindle width is less than 1 μ between the daughter masses of chromatin (Plate 35, figs. 4 and 5) although Hertwig (1889) reported that there is a middle part of the telophase spindle which is wider. The two micronuclear divisions are almost complete before the fission furrow is formed and one daughter micronucleus from each mitosis passes into each daughter paramecium.

By a series of binary fissions and provided that the animals are prevented from undergoing autogamy, a single paramecium can give rise to a population of individuals which is called a 'clone'. There is conclusive evidence that each animal of a clone is genetically identical and this is evidence in favour of the normality of the micronuclear mitosis.

Ehret and de Haller (1963) observed a reorganisation of structures in the macronucleus of *P. bursaria* immediately before fission. The 'large bodies' ceased to be in grape-like bunches and became dispersed singly, while the small bodies became more finely dispersed and then returned to their former state. The significance of these changes is unknown and there is no evidence for their occurrence in *P. aurelia*.

The macronucleus appears to divide amitotically without breakdown of the nuclear membrane (Plate 37). Immediately before fission the macronucleus occupies a central position, which is dorsally away from the gullet (Hertwig, 1889). As the fission furrow begins to deepen the macronucleus elongates and then constricts in the middle to form a shape like an hour-glass (Plate 37, figs. 1 and 2). It finally separates into two equal halves when the furrow is completed.

Carasso and Favard (1965) showed that in certain peritrichous ciliates there are large numbers of microtubules within both the macronucleus and the micronucleus at the time of nuclear division. The microtubules constituted intranuclear spindles. In the dividing macronucleus there appeared to be a number of spindles each separate from each other. For *Paramecium*, the only macronuclear microtubules to have been recorded seem to be the rather peculiarly disorganised ones reported by Vivier and André (1961) in *P. caudatum*. In sections of *P. aurelia*, not in fission, we have in fact once observed microtubules in the macronucleus (Plate 33, fig. 3); they have also been found in a small proportion of isolated macronuclei. On the other hand the amitotically dividing macronucleus contains many microtubules, particularly in the region of the isthmus (Plate 37, figs. 3 and 4).

As described in the early account of the fission process given by Hertwig (1889), the gullet of the parent paramecium passes to the anterior daughter animal; the gullet of the posterior daughter animal was said to arise by being budded off from the parent structures. More recently the development of the new gullet in *P. aurelia* has been studied by Roque (1956a), Yusa (1957) and Porter (1960). Ehret and Powers (1959) studied gullet development in *P. bursaria* under the light microscope while Ehret and de Haller (1963) used the electron microscope. From the observations of Ehret and de Haller (1963) it may be inferred that the development of the new gullet is initiated in *P. bursaria* about 40 min before the fission furrow can be observed. There is then a proliferation of basal granules in the right wall of the vestibular region of the pellicle adjacent to the posterior end of the endoral membrane. These will later become the basal granules of the new peniculus and quadrulus. Roque (1956b) suggested that the basal granules of the new gullet are derived in some way from those of the endoral membrane of the old gullet but this view has not been confirmed. The ribbons of basal granules in the new peniculus and quadrulus progressively increase in length and become ciliated on the cell surface to the right of the old gullet; this process continues until fission takes place. The electron micrographs of Haller, Ehret and Naef (1961) and Ehret and de Haller (1963) clearly demonstrate the relative positions of the old gullet and the immature new gullet. Invagination of the new gullet occurs just before and during fission. The fission furrow passes between the old and new gullets and separates them.

Gillies and Hanson (1967) draw a distinction between stomatogenesis by budding as has been described for *P. aurelia* and stomatogenesis by surface invagination which they describe for *P. trichium*. These authors point out that the vestibulum of *P. trichium* is not as deep as the vestibulum in *P. aurelia* and this may cause differences in the appearance of stomatogenesis. When, however, a detailed comparison is made between the various accounts of stomatogenesis given for *P. aurelia*, *P. trichium* and *P. bursaria*, the similarities are very marked. In all the species the new gullet originates as a proliferation of basal granules at the right posterior margin of the buccal overture between the endoral membrane and the first vestibular kinety. At a later stage the basal granules are organised into rows of granules which will form the new peniculus and quadrulus. In all species invagination plays an essential part in the formation of the new gullet.

It will be recalled that the dorsal and ventral peniculus and the quadrulus each consists of four rows of basal granules with their cilia. During the development of a new gullet anlagen, however, Ehret and de Haller (1963) observed stages in which two rows of granules were ciliated and two were unciliated. They estimated that the unciliated basal granules were no more than 10 min old and that in another 10 min they would become ciliated. New basal granules and new cilia always developed on that side of existing ciliated basal granules which was in the direction of the old endoral membrane and ribbed wall. Ehret and de Haller therefore searched for precursors of basal granules with the advantage of knowing where to look. They identified precursor bodies as small circular vesicles about 100–200 mμ in diameter, but the cytoplasm in these regions appears so generally populated with vesicles that the identification cannot be regarded as certain.

Several light microscopists including Gillies and Hanson (1967) have described the first stage in the development of a new gullet as an anarchic field of basal granules which only later become organised into rows. Hufnagel (1967) used phase and electron microscopy to photograph the anarchic field in preparations of isolated pellicle from *P. aurelia*. There are disordered basal granules in a strip of cell surface about 2 μ wide between the endoral membrane which borders the ribbed wall on one side of the field and the first kineties of the pellicle on the other side. These preparations show no fine detail, however, and Hufnagel considers that the anarchic field is present throughout the interphase period. It is not easy to reconcile a disordered arrangement of basal granules with the idea of the synthesis of new basal granules immediately anterior to an ordered pattern of pre-existing basal granules. The basal granules could perhaps be derived from precursor bodies which are themselves synthesised in contact with mature basal granules, in the same way that basal granules were observed to proliferate in mammalian cells by Dirksen and Crocker (1965).

Immediately before and during furrow formation, Ehret and de Haller (1963) examined the pellicle system at the cell surface in the furrow region and noted that a new basal granule always appears, just in front of an existing basal granule. Where there was a pair of basal granules at the centre of a corpuscular unit of pellicle, two basal granules formed anteriorly to each unit of the system (Plate 38, figs. 1 and 2). Immediately before the formation of new basal granules, Ehret and de Haller observed an increase in the numbers of certain mitochondria which were modified to include a vacuolated space within their outer membranes. These mitochondria were mostly in the subpellicular region of the cell and frequently contained circular or spherical vesicles similar to those which Ehret and de Haller suggested might be precursor bodies for basal granules. The identification of precursor bodies is a difficult and controversial task but the observations of Ehret and de Haller do provide strong evidence against the idea (once strongly held) that new basal granules arise by direct division of mature basal granules. There is no evidence for the division of basal granules.

Dippel (1965), working with *P. aurelia*, studied the duplication at fission of ciliary units of the pellicle where there were initially two basal granules. New basal granules formed to the anterior and in contact with the existing ones, so that a row of four basal granules was formed which were alternately new and old. Membrane growth and partition then occurred to separate the four basal granules into two pairs, each with one old and one new basal granule at the centre of a partly new unit of the pellicle. In the posterior unit the old posterior basal granule retained its kinetodesmal fibre and parasomal sac, while in the anterior unit the old posterior basal granule acquired a new kinetodesmal fibre and a new parasomal sac. The interpretation which Gillies and Hanson (1968) put upon their light microscope observations with *P. trichium* represents an extension of this scheme; they consider that up to four more new basal granules may immediately form just in front of each of the four basal granules in the row produced according to Dippel. This row of eight basal granules will then extend longitudinally to give four pairs. Parasomal sacs and kinetodesmal fibrils will subsequently form as before and membranes develop so that one original corpuscular unit of the kinety may form four.

Our observations on the pellicle during the period of rapid growth of the kineties at the fission stage (Plate 38) show the basal granules in pairs; the younger basal granule of each pair may be recognised because it is not in contact with the cell surface (Plate 38, figs. 1 and 3). According to Dippell (1968) new basal granules are perpendicular to old basal granules at the time of their formation and subsequently reorient until they are parallel. The rapid growth of both the plasma membrane and the alveolar membranes appears to increase progressively the separation between adjacent pairs of basal granules (Plate 38, figs. 3 and 4). During this period there are bundles of microtubules which run in the cortical ridges between adjacent kineties (Ehret, Alblinger and Savage, 1964; Plate 36, fig. 2 and Plate 38, fig. 6). The growth is also associated with the presence of many vesicles, between 50 mµ and 300 mµ in diameter (Plate 38, figs. 4 and 5) which are just below the cell surface in the ridges; many microvilli were observed at the cell surface itself (Plate 38, fig. 7).

It would appear that during successive fission cycles the kineties continually undergo longitudinal extension as new basal granules form to the anterior of old ones and then development takes place within the new kinetosomal territories. This growth process is not, however, uniform along the length of a kinety. In an animal about to divide the rate of growth is higher in the equatorial region, which will become both the anterior of the opisthe and posterior of the proter; and the rate is correspondingly lower in the polar zones. The scheme of Gillies and Hanson (1968) allows equatorial units to triplicate and quadruplicate while polar units remain undivided and the intermediate units duplicate. Sonneborn (1963) has suggested that there is a gradient of growth rate which is high near the equator of an animal about to divide and low towards both poles. Later, as the furrow forms, Gillies and Hanson (1968) consider that the region of higher growth rate moves to the equatorial planes of the proter and opisthe.

Beisson and Sonneborn (1965) studied the behaviour, during successive fissions, of certain abnormal individuals of *P. aurelia* into which a few abnormal kineties had been grafted in an inverted orientation so that the kinetodesmata ran to the posterior rather than to the anterior end of the animal. Inverted kineties were retained in their inverted orientation for many fissions, showing that new kinetosomal territories are organised according to the orientation of the kinety to which they belong rather than according to the orientation of the paramecium. During the course of many fissions, Beisson and Sonneborn (1965) also noticed a lateral shift which displaced the abnormal kineties until they were predominantly on the cell's left side. The mechanism of this lateral movement is not understood but it has been considered from a theoretical point of view by Ehret (1967). Cases of reversed kineties similar to those studied by Beisson and Sonneborn were found by Gillies and Hanson (1967) to occur naturally in about 2 per cent of individuals from a mass culture of *P. trichium*.

During the past eight years an effort has been made to collect quantitative data on parameters related to growth and development at measured intervals during the fission cycle of several species of *Paramecium* (see Diagram 11). Kimball, Caspersson, Svenson and Carlson (1959) showed that when an unlimited food supply was present, individuals of *P. aurelia* showed an exponential increase in dry weight with time throughout the interfission interval. The measurements of Ehret and de Haller (1963) showed a continuous increase in cell length throughout the interfission interval for *P. bursaria*. Several spectrophotometric measurements have been made of DNA increase in the macronucleus and the micronucleus. Rao and Prescott (1966) made measurements for *P. caudatum*. For *P. aurelia*, Kimball and Barka (1959) and Kimball and Perdue (1962) made measurements on the macronucleus and Woodward, Gelber and Swift (1961) on the micronucleus. During the first two hours after fission there is no detectable DNA synthesis but DNA synthesis in the macronucleus occupies most of the second half of the interfission interval. According to Raikov, Cheissin and Buze (1963), DNA synthesis of *P. caudatum* can in some cases begin earlier in the micronuclei than in the macronucleus while in others it begins later and there appears to be no strict correlation between the times of synthesis in the two kinds of nuclei.

60

Another developmental problem concerns the origin and growth of trichocysts during the fission cycle. Mitrophanow (1905) seems to have been the first to recognise the endoplasmic origin of trichocysts in *Paramecium*. Electron-microscope studies on the epigenesis of trichocysts have been conducted by Yusa (1963) on *P. caudatum*, by Ehret and de Haller (1963) on

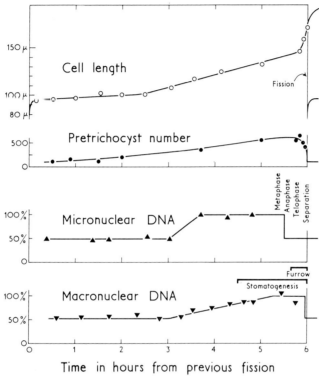

DIAGRAM 11 Growth and fission cycle of *Paramecium*. Graphical data on cell length and pretrichocyst number are from Ehret and de Haller (1963) for *P. bursaria*, on micronuclear DNA from Woodward and Swift (1961) for *P. aurelia*, and macronuclear DNA from Kimball and Barka (1959) for *P. aurelia*. *P. bursaria* has a 12-hour fission cycle but all data have been fitted to the 6-hour fission cycle of *P. aurelia*.

P. bursaria and by Yusa (1965) on the ciliate *Frontonia vesiculosa*. The genus *Frontonia* bears a close phylogenetic relationship to the genus *Paramecium*. Yusa (1963, 1965) observed the regeneration of trichocysts after they had been made to discharge by an electrical shock treatment but Ehret and de Haller observed their regeneration during a normal fission cycle. In both cases the trichocysts were observed to arise by an orderly developmental process from pretrichocysts in the endoplasm of the cell and in no case was there any involvement of basal granules or parent organelles. Our own observations on trichocyst development in *P. aurelia* are in agreement with this view (Plates 42 and 43). Trichocysts develop from an amorphous mass of fine disoriented filamentous elements contained within a membrane which limits an irregularly rounded vesicle about 0.1 to 0.3 μ in diameter. The vesicle enlarges and closely packed fibrillar elements condense within it to form a dense oval body which contains regular crystalline periodicities which measure between 8 mμ and 15 mμ in different sections (Plate 42, figs. 1 and 3). A single row of dense granules of about 13 mμ diameter lies outside the crystalline body of an immature trichocyst (Plate 43). The tip of the trichocyst first forms as a dense object in continuity with the body (Plate 42, figs. 4 and 5; Plate 43, fig. 2) and with similar periodicities. In maturity the tip of the trichocyst retains its periodicity (Plate 12, fig. 1) while the body of the mature trichocyst appears structureless (Plate 12, fig. 1; Plate 41, figs. 1–5). This apparent change in ultrastructure with maturity was observed by Yusa (1964) for *P. caudatum*. In *P. aurelia* maturity is reached for most trichocysts in one hour after fission. The periodicity displayed within the shaft which is formed when a trichocyst is discharged (Plate 13, figs. 2, 3 and 6) is quite different from the periodicity of an immature trichocyst.

Pretrichocyst stages are found anywhere in the endoplasm, in positions which seem to bear no relationship to the sites in the pellicle that they will eventually occupy in maturity. Only when the pretrichocysts have grown a tip and sheath to become juvenile trichocysts do they occupy their final positions at the cell surface where the change to maturity is made.

Ehret and de Haller (1963) from morphological studies and Ehret, Savage and Schuster (1966) from experiments with tritiated leucine, estimated that trichocysts of *P. bursaria* can develop from pretrichocyst stages in less than 30 min. The experiments of Ehret, Savage and Schuster (1965) with tritiated thymidine also suggest that trichocysts may contain some DNA and RNA as well as protein.

The obvious difference in electron density of the bodies of mature and immature trichocysts in *P. aurelia* leads to a ready appreciation of their relative abundance (Plate 41, figs. 2 and 5). In *P. aurelia* our own observations show there is a corresponding difference in affinity for the dye toluidine blue after osmic fixation which enables mature and immature trichocysts to be readily distinguished by light microscopy (Plate 41, figs. 3, 4, 6, and 7). During interfission stages, immature trichocysts are commonly only about 5 per cent of the total but during fission immature trichocysts are very numerous, as Ehret and de Haller showed for *P. bursaria*. These authors do not report on the abundance at fission of old trichocysts which have remained in maturity during the preceding interfission period. In *P. aurelia* at fission nearly all the trichocysts are newly synthesised and the old mature trichocysts are located only at the extreme ends of the animals, remote from the fission furrow (Plate 41, figs. 5 and 6). This is a surprising observation because it suggests that most mature trichocysts are discharged or degenerate before fission, whereas previously they had been thought to be conserved. The work of Ehret, Savage and Alblinger (1964) supports the idea that mature trichocysts are conserved at fission because they found that isotopically labelled trichocysts retained their label without dilution for at least three fission generations. Our observations suggest that in *P. aurelia* this may be true only for trichocysts located at the two poles of the animals.

11 Conjugation

The taxonomic species *Paramecium aurelia* is subdivided into at least fourteen numbered varieties or syngens as they are now called (Sonneborn, 1958) each of which is effectively a genetic species. Within each syngen there are no more than two mating types and when animals of the same syngen are mixed (under the appropriate environmental conditions described by Sonneborn, 1950) they may form clumps from which they subsequently emerge as conjugated pairs, each pair consisting of one member from each of the two mating types. Mating types were first discovered in *P. aurelia* by Sonneborn (1938b and c). Subsequently they have been found in other ciliates.

The behaviour of the nuclei during the conjugation of *P. aurelia* has been described by many authors. Descriptions were given by Hertwig (1889) and

DIAGRAM 12   The diagrams show the nuclear events during a normal conjugation and up to the first postconjugal fission in *P. aurelia*. They are based upon Beale (1954), Grell (1967), Jones (1956), Kościuszko (1965) and other authors. An account of the conjugation process is given in the text and the times and durations of the micronuclear, macronuclear and cytoplasmic events are given in Diagram 13. In stage 1, each conjugant has one large macronucleus and two small micronuclei. For stages 1 to 8, macronuclei can be distinguished from micronuclei by size. In stage 9 and subsequently, the fragments of the old macronucleus are recognisable by the dotted shading which resembles that of the macronucleus. The third conjugal division of the micronuclei produces a stationary gamete nucleus (St) and a migratory gamete nucleus (Mg) in each conjugant and stage 10 shows reciprocal nuclear exchange between the migratory nuclei. Stage 11 shows a synkaryon (Sy) in each conjugant which results from the fusion of the stationary nucleus and the migratory nucleus from the other conjugant. Stage 14 shows the oral space (OS) from which macronuclear fragments are excluded after the first postconjugal mitosis (see also Plate 45, fig. 3). In stages 16 and 17 are seen the macronuclear anlagen (Ma) in growth, the micronuclei (Mi) and the old fragments.

64

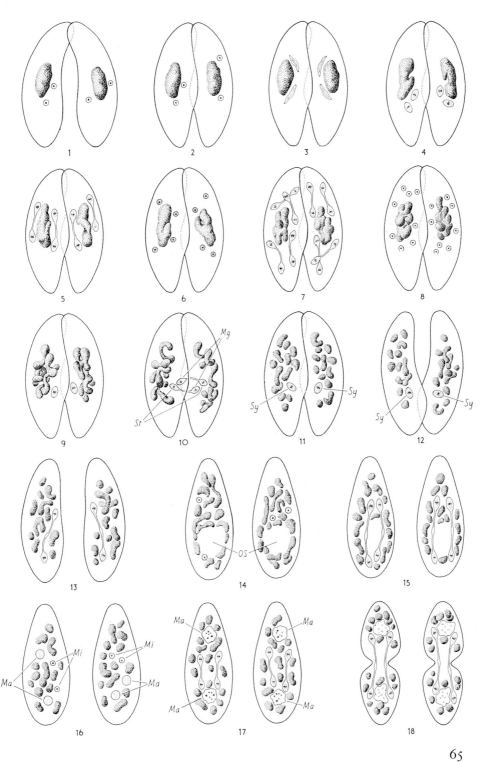

Maupas (1889) and among more recent descriptions are those of Beale (1954) Grell (1967), Sonneborn (1947, 1954), Kimball and Gaither (1955), Jones (1956) and Kościuszko (1965). The latter two accounts are illustrated with photographs taken by light microscopy (see also Plates 44 and 45). The drawings of conjugation stages (Diagram 12) are based on a consideration of the published accounts. Electron-microscope studies have been undertaken to investigate the nature of the surface of contact between pairs in conjugation (Vivier and André, 1961; André and Vivier, 1962; Schneider, 1963; and Inaba, Suganuma and Imamoto, 1966), but ultrastructural studies on most other aspects of conjugation have not been made except for the development of macronuclear anlagen by Jurand, Beale and Young (1964). The available literature is unhelpful if information is required on the duration of the various phases of conjugation and therefore the table (Diagram 13) has been compiled to serve as a rough guide. The tabulated estimates of stage times are based on figures that have been quoted by many authors. It must be emphasised, however, that stocks of *P. aurelia* differ in the times they take to pass through similar stages of development.

When conjugation is induced by mixing paramecia of complementary mating types, the individual pairs are first attached along their ventral surfaces at their anterior ends (hold-fast union) but soon afterwards they become attached in the region of their gullets also (paroral cone union). Mijake (1966) found that a local disappearance of cilia must occur before hold-fast union can take place in *P. multimicronucleatum*. He also found that this local deciliation could be induced by the introduction of cilia from animals of complementary mating type or with certain chemicals. In conjugation the deciliation allows the outer pellicular membranes of each conjugant to become apposed and parallel to each other along the ridges of the pellicle between the rows of basal granules (Plates 46–48). At intervals these membranes fuse and dissolve to form a series of communicating cytoplasmic bridges or pores (Plate 46, fig. 2; Plate 48, fig. 2). There is much granular electron-dense material in the cytoplasm surrounding the pores (Schneider, 1963) which Inaba, Suganuma and Imamoto (1966) suggest may represent dissolution products from the membranes of the pellicle. Soon after the communicating pores have formed along the tops of the

66

CONJUGATION IN *P. AURELIA* AT 25° C

The approximate time and duration of the stages

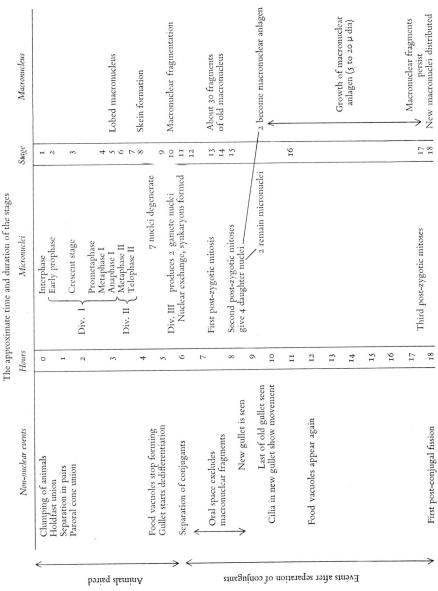

Non-nuclear events	Hours	Micronuclei	Stage	Macronucleus
Clumping of animals	0	Interphase	1	
Holdfast union	1	Early prophase	2	
Separation in pairs				
Paroral cone union	2	Crescent stage (Div. I)	3	
	3	Prometaphase	4	Lobed macronucleus
		Metaphase I	5	
		Anaphase I	6	Skein formation
		Metaphase II (Div. II)	7	
		Telophase II	8	
Food vacuoles stop forming	4	7 nuclei degenerate		
Gullet starts dedifferentiation	5	Div. III produces 2 gamete nuclei	9	Macronuclear fragmentation
			10	
Separation of conjugants	6	Nuclear exchange, synkaryons formed	11	
			12	
Oral space excludes macronuclear fragments	7	First post-zygotic mitosis	13	About 30 fragments of old macronucleus
			14	
New gullet is seen	8	Second post-zygotic mitoses give 4 daughter nuclei	15	
Last of old gullet seen	9	2 remain micronuclei		2 become macronuclear anlagen
Cilia in new gullet show movement	10			
	11			
Food vacuoles appear again	12			
	13		16	Growth of macronuclear anlagen (5 to 20 μ dia)
	14			
	15			
	16			
	17	Third post-zygotic mitoses	17	Macronuclear fragments persist
First post-conjugal fission	18		18	New macronuclei distributed

Animals paired

Events after separation of conjugants

DIAGRAM 13 The main nuclear and cytoplasmic events which occur during a normal conjugation are tabulated against a time-scale in hours. The stage numbers of Diagram 12 are also given against the same time-scale. The main text gives the sources of the published information presented here in tabular form. Different stocks of *P. aurelia* pass through particular stages of conjugation at different rates, so that the main function of the table is to give general guidance concerning the relative durations of the cytological events.

ectoplasmic ridges the communicating area becomes flattened. The region of contact appears devoid of cilia or trichocysts but the basal granules are reported to remain intact and kinetodesmal fibrils have been observed (Plate 47, figs. 1 and 3).

Vivier and André (1961) cut serial sections of conjugating pairs and showed that in the paroral region the conjugants are in contact along the surface just to the right of the gullet opening. Normally the conjugants finally separate after about 5–6 h; the paroral attachment is always the last to disappear and in abnormal cases, where separation never occurs, double animals are formed which are held together in this region.

Three micronuclear divisions take place while the conjugants are paired (Diagram 13). The first of these divisions occupies about two-thirds of the time the animals are in contact with each other. It has a specially modified prophase of long duration and is almost certainly to be identified with the first meiotic division. That segregation of genes occurs during conjugation has been firmly established as a result of many genetic studies (e.g. Beale, 1954). Earliest prophase of the first conjugal division is identified when the normal ring-shape of the chromatic area in interphase is replaced by a more uniform distribution of the stainable material (Plate 44, fig. 2). At a slightly later stage, two small Feulgen-positive bodies, each about $1\ \mu$ in diameter (Plate 44, fig. 3) can be distinguished from the uniformly stained background of the rest of the nucleus (Jones, 1956). These bodies may represent heteropycnotic chromosomes. At later stages the prophase nucleus shows a remarkable elongation to about $20\ \mu$ accompanied by an increase in volume to give a crescent stage (Plate 44, figs. 4 and 5) which bears some points of resemblance to the bouquet stage characteristic of early meiotic prophase in certain metazoa. During the crescent stage several long threads of chromatin can be seen stretched out parallel to each other in the direction of the greatest elongation of the nucleus. At the end of the crescent stage the nucleus becomes more rounded and four rather than two heteropycnotic bodies have been observed (Jones, 1956).

A short prometaphase is followed by a metaphase stage during which large numbers of minute chromosomes can be seen; these vary in size from about one micron down to the limits of optical resolution (Plate 44, fig. 7).

Chromosome counts have been made at this stage for a variety of different stocks of *P. aurelia* by Dippell (1954), Jones (1956) and Kościuszko (1965). The diploid number of chromosomes was found to vary between 74 and 126 which, even when allowance is made for the errors of counting, still suggests that different stocks of *P. aurelia* possess different numbers of chromosomes. It is possible, however, that some of the small pieces of chromatin which have been counted as individual chromosomes in these studies, may in fact have been linked in a common chromosome by a chromatin thread which was below the resolution of the microscope. In this case a true estimate of chromosome number might be much less than that quoted.

After completion of the anaphase in the first meiotic division there is no interphase and the second division takes place immediately. When this division is complete there are eight haploid telophase nuclei in each of the two conjugants (Plate 45, fig. 1). Normally, seven of these nuclei then begin to degenerate so that by the time conjugation is complete they cannot be detected. Occasionally another mitosis precedes the nuclear degeneration (Preer, 1967). The eighth nucleus, on the other hand, does not degenerate but it is always observed to occupy the paroral region of the cytoplasm to the right of the gullet in each conjugant. Sonneborn (1955) has suggested that this region of cytoplasm plays a protective rôle towards the nucleus, because all eight micronuclei degenerate in a certain stock if none of them enters the paroral region.

Only the haploid micronucleus in the paroral region undergoes the final nuclear division in each of the conjugants. Genetic studies have proved that this division is equatorial (Jollos, 1921; Sonneborn, 1947; Beale, 1954). As a result of this division two functional gamete nuclei are formed in each conjugant. Light microscopy has shown (Jones, 1956) that during the telophase separation of the daughter gamete nuclei, one of the nuclei (termed the migratory or male gamete nucleus) is pushed into the cytoplasm of its mate (stage 10 in Diagram 12 and Plate 45, fig. 2) where fusion of the gamete nuclei in each animal quickly follows to form a synkaryon (stage 11 in Diagram 12). In studies employing electron microscopy (André and Vivier, 1962; Vivier, 1965, with *P. caudatum* and Inaba, Suganuma and Imamoto,

1966, with *P. multimicronucleatum*) the migratory pronucleus was observed just before it passed through one of the cytoplasmic pores into the cytoplasm of the mate. In these cases no special apparatus was seen which might suggest a mechanism for the migratory movement but the nuclear membrane was highly convoluted so that the nucleus resembled an amoeba. It was also clear that the nucleus must be constricted to let it pass through one of the narrow cytoplasmic bridges. Genetic analysis (Jennings, 1913; Sonneborn, 1947; Beale, 1954) has confirmed that reciprocal cross-fertilisation must occur so that the synkaryons in each conjugant are genetically equal. Normally the conjugants separate immediately the synkaryons are formed, and no appreciable exchange of cytoplasm between the conjugants takes place. Many examples, however, of a long paroral union have been studied and in these there is evidence of a considerable cytoplasmic exchange (Sonneborn, 1947; Beale, 1954).

While the micronuclei of the conjugants are in the metaphase of the first meiotic division (stage 4 in Diagram 12) the macronucleus begins to show signs of change. Its outline appears more irregular and contains lobes and grooves. By the time the second meiotic division has been completed, the irregularities are much more obvious as the nucleus reforms into long skeins of Feulgen-positive material (stage 8 in Diagram 12). At the synkaryon stage the macronucleus has begun to break up into smaller pieces and this process continues so that smaller and more numerous fragments are produced. These fragments have been shown to persist through the two fissions following conjugation. At this stage, Jurand, Beale and Young (1964) observed them in sectioned material under the electron microscope but there were no signs of degeneration, which must therefore occur at a later stage.

Between the separation of the conjugants and the occurrence of their first postconjugal fission (a period of about 12 h) three mitotic divisions take place (Diagram 13). The first two occur quite soon after the formation of the synkaryon and as a result four diploid nuclei are formed in each conjugant (stage 14 in Diagram 12). After the first postzygotic mitosis the two daughter nuclei may both be in either the anterior or posterior half of the cell or there may be one of them in each half (Jones, 1956). For a period after the second mitosis there are two nuclei in each half of the cell. Two of these four nuclei

70

normally enlarge and differentiate into macronuclear anlagen. Jones (1956) was able to distinguish macronuclear anlagen from micronuclei while two nuclei were still towards each end of the cell; he obtained photographs showing the two anlagen in the anterior region and in another case a macronuclear anlagen together with a micronucleus towards each pole of the cell. Later the nuclei are located near the centre of the cell. The macronuclear anlagen increase in size endomitotically so that they have almost reached full macronuclear size at the time of the first fission (stage 18 in Diagram 12). The nuclei which do not become macronuclear anlagen become micronuclei and they divide mitotically immediately before the first postconjugal fission. The two macronuclear anlagen do not divide at this fission but become distributed so that each animal receives one. Subsequent fissions are normal.

Sonneborn (1958) studied the differentiation of new macronuclear anlagen using abnormal stocks of *P. aurelia* and concluded that a cytoplasmic factor must be responsible. It was also demonstrated that centrifugation at the time of the second postconjugal division could modify the normal 2:2 ratio of macronuclei to micronuclei. Jones (1956) showed that immediately before the second postconjugal mitosis a special clear space could be recognised in the endoplasm of the oral region (Plate 45, fig. 3) because this was the only part of the cytoplasm which had no nuclei and no old macronuclear fragments, there being about thirty of these fragments in the cytoplasm of each paramecium at this stage. Because the telophase spindles of this second division extend at least half the length of the animal, it seems that two telophase nuclei must pass through this special oral region of the cytoplasm; Jones (1956) suggested that these nuclei must eventually give rise to macronuclei. Subsequently the macronuclear anlagen have been shown to contain vacuolar regions each of which possesses a Feulgen-positive granule (Plate 45, fig. 4 and Jones, 1956; Egelhaaf, 1955; Jurand, Beale and Young, 1964) but the significance of these observations is unknown.

Sonneborn (1939) also worked with *P. aurelia* from stock R of syngen 1 and found that after conjugation 20 per cent of the animals possessed more than two macronuclear anlagen: in most cases they possessed three or four but occasionally there were up to ten anlagen. In these cases the anlagen were distributed rather than divided in the early postconjugal fissions until each animal had only a single macronucleus.

At a stage about half way between the separation of the exconjugants and the first postzygotic fission (and at a corresponding stage following autogamy) the two macronuclear anlagen, which are then about 10 μ in diameter (Plate 49, figs. 3 and 4) have been observed to approach each other very closely near the centre of the cell so that they are separated by a layer of cytoplasm less than 0.5 μ in width. Maupas (1889) and Hertwig (1889) sketched what appears to be a similar stage but its significance still remains unknown. Normally the anlagen separate again, but Sonneborn (1947) and Nanney (1956) reported that under conditions of starvation the two anlagen in the exconjugants became irreversibly fused to form one macronucleus.

Jurand, Beale and Young (1964) studied the normal development of macronuclear anlagen using electron microscopy. The young anlagen contained neither large nor small bodies but only fine fibrigranular material of low density (Plate 49, fig. 1). The first dense bodies were observed when the anlagen were of diameter between 10 and 13 μ, and these bodies (Plate 49, fig. 2) were usually larger than the large bodies in mature macronuclei. At a later stage the dense material appeared in the form of sponge-like aggregates (Plate 49, fig. 5). Electron-dense small bodies were observed in new macronuclei after the first postconjugal fission had occurred (Plate 49, fig. 6) but at this stage these small bodies were slightly smaller than those of old macronuclear fragments in the same animals. This observation is consistent with the idea that small bodies originate by condensation from the fine Feulgen-positive material. In animals of a similar stage, there are large regions of irregular electron-dense material (Plate 49, fig. 7) which was shown to be rich in RNA. The disintegration of these regions was considered to provide material for the large bodies of the mature macronucleus.

During conjugation the animals swim about all the time and at first the contractile vacuoles continue to function normally (Wichterman, 1953). After the two meiotic divisions, however, food vacuoles are no longer formed and the gullet begins to dedifferentiate (Porter, 1960). Irregular masses of pyronin-positive material have been seen in the old gullet at the time of nuclear exchange (Jones, 1956). The new gullet was first observed by Hanson and Gillies (1956) as a small protuberance on the inner right posterior edge of the buccal opening, at the time of the first postconjugal mitosis. The old gullet, therefore, seems to degenerate *in situ* and the new gullet takes over the site of the old buccal overture. The last remnant of the old gullet is seen after the second postconjugal mitosis, and food vacuole formation recommences when the new macronuclear anlagen may be easily distinguished. There is no reason to suppose that the new gullet forms in a dissimilar manner to the formation of a new gullet at fission, although the point has not been closely investigated; but degeneration of the old gullet only takes place in conjugation and autogamy.

Roque (1956a) studied the degeneration of the old gullet and the development of the new gullet by making light microscope observations on *P. aurelia* during autogamy. This author points out that it is easier to make observations on the gullet of a paramecium in autogamy rather than in conjugation because in autogamy there is no pairing partner to obscure the view in a whole-mount preparation. Roque (1956a) states that for autogamy the earliest development of the new gullet begins at the end of prophase of the first meiotic division. Roque (1956b) also claims that the development of the new gullet is similar in fission, conjugation and autogamy.

12 Cytoplasmic symbiotic inclusions

Certain stocks of *Paramecium aurelia* produce and exude into the surrounding water a poisonous agent which can kill other sensitive paramecia. The first killer stock was discovered by Sonneborn (1938, 1939) who noticed that sometimes when a mixed population of two genetically different stocks is kept in the same culture medium, the animals of one stock die out while the others survive. There are several types of killer stock which differ according to the behaviour of the sensitive paramecia when they are affected by the substance produced by killer paramecia in the same medium (Preer, Siegel and Stark, 1953). Some killers cause sensitives to spin in a clockwise direction, some cause vacuolisation and swelling of the cytoplasm, while others induce paralysis alone. The sensitives may die in each case. Shortly after this discovery Sonneborn was able to show that the killing property of a stock is inherited but that the genetic factors must reside in the cytoplasm. He called these factors 'kappa particles'. Furthermore, it became apparent that besides the kappa factors present in the cytoplasm a dominant gene K must also be present in the nucleus if the kappa particles are to be maintained (Sonneborn, 1959).

There are also other groups of killer paramecia which do not kill sensitive animals when they are present in the same medium but only when they are allowed to conjugate with them. In such cases either the sensitive excon-jugant dies immediately after conjugation or the daughter cells die after a few fissions. To these strains the term mu or 'mate killer' is applied.

In all types of killer paramecia there are certain inclusions which may be observed in the cytoplasm by light microscopy (Plate 50) and which presumably are responsible for the production of the substance which is poisonous to other paramecia. In the case of the kappa type of killer paramecia, the lethal substance released into the medium is a particulate material originally called 'paramecin', which van Wagtendonk (1948) showed must contain both protein and DNA. Later work of Preer and Stark (1953), Smith (1961) and Mueller (1963) identified this particulate material with the B type of kappa particle which is derived from a self-reproducing N particle. Dryl and Preer (1967) observed that sensitive animals seemed unresponsive to kappa during cell fission and during those phases of conjugation and auto-gamy when no food vacuoles are formed. The B type particle, therefore, probably delivers its toxin to sensitive animals after first being taken into the food vacuoles.

In mate-killers the nature of the poisonous factor is unknown. The presence, however, of the cytoplasmic inclusions in mate-killer, as in kappa, strains is also dependent upon the presence of certain specific nuclear genes (Siegel, 1953; Gibson and Beale, 1961; Beale and Jurand, 1966).

There are certain other types of killer paramecia. One of these is the lambda type which behaves similarly to kappa but whose killing action is more rapid. The cytoplasmic inclusions kappa, mu and lambda have been studied by light microscopy, cytochemistry and electron microscopy. All of them appear to resemble bacteria. From the general biological point of view it is probably correct to describe the relationship between these particles and their host strains of paramecia as one of well-adapted endosymbiosis (Buchner, 1953). The particles must derive nutrients from the host while a killer stock of paramecium has an advantage over other stocks when com-peting for a restricted food supply.

Certain stocks of *P. aurelia* have been found to contain non-killing symbionts which also resemble bacteria. One example is the case of the pi particles studied by Hanson (1953, 1954) which may be regarded as a mutant form of kappa, because they are dependent upon the gene K for survival; the presence of pi, however, does not confer protection against kappa. Pi particles have similar staining properties to kappa but unlike kappa they never contain refractile inclusions. Another example of a non-killer symbiont in *P. aurelia* is nu, which was studied by Sonneborn, Mueller and Schneller (1959).

All the symbiotic particles of *P. aurelia* grow by elongation and divide

75

into two equal daughter bodies by the formation of a constriction furrow as in bacteria (Plate 51, fig. 4; Plate 52, fig. 4; Plate 54, fig. 5). If the fission process in the host paramecium is inhibited by a period of starvation, the concentration of particles in the cytoplasm is considerably increased. Conversely the concentration of particles is diminished if the paramecia are given a period of intensive feeding.

Kappa particles were shown by Preer (1950) to contain Feulgen-positive material. Smith and van Wagtendonk (1962) analysed them chemically and showed that they contain both DNA and RNA. They have been examined under the electron microscope by Dippel (1958) and Hamilton and Gettner (1958). At this level it is easy to distinguish between two types of kappa particle which are usually present in the cytoplasm of the same host. One, called the N type (Plate 51, fig. 4), is like a rod-shaped bacterium and is 3 μ long and 0.5 μ wide. The other type is more rounded, about 0.5–1 μ in diameter, and usually has one or occasionally two or even more refractile lamellar inclusions (Plate 51). These kappa particles were named 'brights' or type B (Plate 51, figs. 1 and 3) because the refractile body within their cytoplasm appears bright when observed by positive phase-contrast microscopy (Preer and Stark, 1953). It is now thought that the N and B types of kappa particle represent two stages in the life cycle of the same symbiotic micro-organism, the N type being the reproductive stage which cannot produce lethal effects whereas the type B particles liberate the killing agent but cannot reproduce (Preer, 1967). Preer and Preer (1967) and Preer and Jurand (1968) also observed virus-like bodies in association with the refractile inclusions in the B type of kappa particle (Plate 51, fig. 7).

Electron microscopy of sectioned material shows that each refractile inclusion within a B type particle is in the form of a ribbon wound in a cylindrical coil like the spring of a clock (Diagram 14). A typical ribbon is 11 mμ thick, 400 mμ wide and about 10 μ long and is wound eight times round a cylinder of granular cytoplasm which is 275 mμ in diameter. Anderson, Preer, Preer, and Bray (1964) have shown that when the refractile bodies inside the 'brights' or type B particles are isolated using sodium deoxycholate as a lysing agent, they have a strong killing property when brought into contact with sensitive animals. These authors and Preer,

Hufnagel and Preer (1966) also examined the isolated refractile bodies by electron microscopy and confirmed a previous observation of Mueller (1962, 1963) that the cylindrical coils can be made to unwind by a suitable chemical treatment.

DIAGRAM 14   The diagram shows the cylindrical coiled shape of a refractile body as it occurs in a type B kappa particle. There are eight turns of the coil in the diagram which is correct for a typical refractile body in kappa particles of *P. aurelia*, stock 51 (see Plate 53, figs. 5 and 6) but in stock 7 (Plate 51, fig. 7) and stock 562 (Plate 53, fig. 1) there are usually a few more turns in the coil. The form of the edge at the inner and outer ends of the coiled refractile strip also differs between stocks. In stock 7 the outer end has an irregular edge and the inner end has a straight diagonal edge (Preer and Preer, 1967), whereas in stock 51 both ends are straight and diagonal.

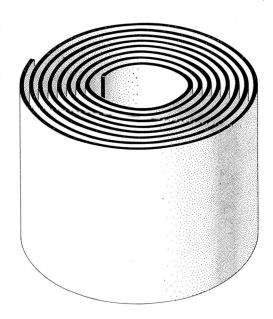

Kappa particles have a plasma membrane about 8 mμ thick and outside it there is another membrane about 10 mμ thick which may correspond to a bacterial cell wall (Plate 51, figs. 4 and 7). The kappa particles contain numerous granules about 10 mμ in diameter. There are also areas of lesser electron density within which run fine threads less than 5 mμ in diameter except in places where they broaden into thicker regions of 15 mμ diameter (Plate 51, figs. 5 and 6). These fine threads are probably DNA.

The mu particles found in mate-killer strains are rather randomly distributed throughout the cytoplasm, although in cases where they are present in very large numbers, they tend to form dense aggregates of many hundreds of particles (Plate 50, fig. 5). They are Feulgen-positive, like kappa particles and the individual particles can be distinguished in paraffin sections using the light microscope (Beale and Jurand, 1960).

In the electron microscope (Plate 54, figs. 4 and 6) mu particles appear to resemble bacteria even more than do kappa particles (Beale and Jurand, 1960, 1966). They are in the form of straight rods, 0.4 μ in diameter and they vary in length between 2 μ and 10 μ. Outside the plasma membrane there is a bacterial cell wall about 19 mμ thick which is seen by electron microscopy as two electron-dense layers each about 7.5 mμ thick separated by a less dense layer about 4 mμ thick (Plate 54, fig. 5). Within the plasma membrane there are many dense granules and fibrillar elements which are distributed irregularly throughout the length of the particle.

In typical bacteria the fibrillar elements which have been identified with DNA usually occur within a compact central region. This difference, however, does not necessarily imply a rejection of the bacterial nature of mu particles because certain bacteria do have such disperse nuclei (Glauert, 1962). Typical bacteria contain mesosomes but the absence of such bodies from kappa, mu and lambda particles may well reflect an adaptation to endo-symbiotic life, because mesosomes are commonly regarded as the bacterial equivalent of mitochondria and mitochondria are absent from certain eukaryotic parasites such as the microsporidian which was studied by Jurand, Simões and Pavan (1967). Stevenson (1967a) demonstrated that mu particles contain compounds such as α, ε diaminopimelic acid and muramic acid which are not known to occur in other than bacterial cell walls.

The third type of endosymbiotic inclusion, the lambda particle, is also rod-shaped like a bacterium (Plate 51, fig. 2 and Plate 52) and has been isolated and chemically analysed by van Wagtendonk and Tanguay (1963). Lambda particles are about 0.7 μ in diameter and so are somewhat thicker than mu particles, and unlike mu particles they do not exceed 4.5 μ in length. Outside the plasma membrane of this particle there is a rather corrugated cell wall about 20 mμ thick (Plate 52, figs. 4 and 6). The cytoplasm of the particle contains granules 10 mμ in diameter but some sections show one or more round areas filled with fine fibrils (Plate 52, figs. 5 and 6). These areas probably correspond to the so-called bacterial nuclei. The bacterial nature of lambda particles is emphasised by the fact that the cell wall of these inclusions is partly covered with bacterial flagella (Jurand and Preer, 1968) which are up to 4 μ long and are uniformly about 25 mμ thick (Plate 51, figs.

1–3). Outside the flagella the lambda particles are separated from the cytoplasm of the host paramecium by a vacuolar membrane about 10 mμ thick.

Even stronger evidence in favour of the bacterial status of the lambda particles arises out of the work of van Wagtendonk, Clark and Godoy (1963) who found that the particles were able to reproduce themselves when grown on a complex, cell-free medium. After having been cultured in this way the lambda particles retained their ability to kill sensitive paramecia and to infect *P. aurelia* of the appropriate genetic constitution.

Sonneborn, Mueller and Schneller (1959) and Schneller (1962) found another type of killer particle in the cytoplasm of *P. aurelia*, syngen 2, which they have called sigma. This resembles lambda in having bacterial flagella and in its rapid killing action but it is distinguished morphologically by its sinuous form and by its length which is from 10–15 μ.

Recently L. B. Preer (unpublished) has used *P. aurelia*, stock 562, to study an endosymbiotic inclusion called alpha (Plate 53, figs. 2 and 5) which appears to be similar to one observed previously by Muller (1856) and Petschenko (1911). It is remarkable in that it is found inside the macronucleus (Plate 50, fig. 4; Plate 53) while the other types have only been observed in the host's cytoplasm. Alpha particles are long and sinuous; they may be 6 μ in length (Plate 50, fig. 4) and have a circular cross-section 0.4 μ in diameter (Plate 53). The alpha particles have three zones of dissimilar cytoplasm: a granular type with bodies which appear to be ribosomes, a dense zone of closely packed granules which are 4 mμ in diameter and a low density zone with only very fine fibrils (Plate 53, fig. 4). The same stock 562 also has cytoplasmic inclusions of the kappa type (Plate 53, fig. 1).

Yet another kind of particle from stocks 214 and 565, both syngen 8, has been observed by Beale, Jurand and Preer (1969). This is called the gamma particle and is characterised by a strong killing action, small size and a tendency to occur in pairs (Plate 50, fig. 3; Plate 54, figs. 1 and 3). Gamma particles from both stocks appear identical in the electron microscope; each particle is ovoid in shape and its longest dimension is only about 0.8 μ; in its cytoplasm is a dense body about 130 mμ in diameter (Plate 54, fig. 3). Pairs of gamma particles have often been observed within the cisternae of the endoplasmic reticulum of the host (Plate 54, figs. 2 and 3).

79

Endosymbiosis is probably of common occurrence throughout the Protozoa but it has been most intensively studied in *P. aurelia*. Preer (1967) lists publications which describe endosymbionts in thirty different stocks of *P. aurelia*; about sixty stocks containing endosymbionts are now known. The terms kappa, lambda, mu, etc., refer to particular classes of endosymbionts. There are many examples of each class in different stocks. These differ in the way sensitive paramecia behave in the presence of the killers (Sonneborn, 1965), or in the detailed structure of the symbionts or in the host genes needed to maintain the symbionts. There is a useful correlation (with some exceptions) between the morphological characteristics of the class of symbiont and its killing action. Thus all known mate-killers have a similar morphology. Refractile bodies are always present in a proportion of the kappas of killer stocks and lambda particles have bacterial flagella but the two groups may also be distinguished by the more rapid lysis produced by lambda stocks.

Plates

Key to Plates

Scales represent 1μ unless indicated

The following abbreviations have been used on the Plates and Diagrams

a	alveolus	mi	micronucleus
am	alveolar membrane	mb	membrane
ap	ampulla	mg	migratory gamete nucleus
as	anterior suture	ml	microtubules
ax	axial granule	ms	microtubular sheet
		mt	mature trichocyst
b	basal granule	mv	microvillus
bo	buccal overture		
br	fibrillar bridge	n	naked ribbed wall
		nt	nephridial tubules
c	nephridial canal		
ci	cilium	o	orthotubular system
co	fibrillar cord	oa	outer alveolar membrane
cp	cytopharyngeal tube	os	oral space
cr	crystals		
		p	parasomal sac
d	granular cortical layer	pe	postesophageal fibres
dc	discharge channel	pf	postciliary tubular fibrils
di	disc-shaped body	pl	plasma membrane
dpn	dorsal peniculus	pn	peniculus
		ps	posterior suture
e	endoral membrane	pt	pretrichocyst
eg	enzyme granules		
er	endoplasmic reticulum	q	quadrulus
f	fine fibrils	r	ribbed wall
fe	ferritin	rb	refractile body
fl	bacterial flagellum	rf	root fibrils
fv	food vacuole		
		s	median septum
g	gullet	sh	sheath of trichocyst tip
gl	glycogen granules	sp	clear space
		st	stationary gamete nucleus
i	injector canal		
ia	inner alveolar membrane	tb	trichocyst body
		tf	transverse tubular fibrils
jt	juvenile trichocyst	tp	terminal plate
		tt	trichocyst tip
k	kinetodesmal fibril		
		v	vestibulum
l	lipid drop	vb	virus-like bodies
		vpn	ventral peniculus
m	mitochondrion		
ma	macronucleus		

G

83

PLATE 1 A general diagram of *Paramecium aurelia*

This is a drawing to show the shape of *P. aurelia* and the relative positions of some of its more important organelles. The gullet opens to the ventral surface and the oral groove is shown as a shallow furrow which runs from the flared vestibulum, round the opening of the gullet anteriorly along the ventral surface towards the anterior pole of the animal. Behind the innermost and narrowest part of the gullet a young food vacuole is seen to have formed. Other food vacuoles are shown in other parts of the endoplasm. The cytopyge through which food vacuoles are ultimately expelled is on the ventral surface posterior to the gullet. The two contractile vacuoles are just below the dorsal surface. The anteriormost contractile vacuole is shown in diastole and the posteriormost contractile vacuole is shown in systole. The two small micronuclei are near the large macronucleus. Cilia are shown beating in waves as for the case of a forward-swimming animal observed in an instant of time. The cilia cover the entire surface of the paramecium although the diagram shows them only over the animal's outline where they are in fact most easily observed. Note the tuft of longer immotile cilia at the posterior pole. Trichocysts, infraciliature and small organelles of the size of a mitochondrion or smaller are not shown in this diagram.

ANTERIOR

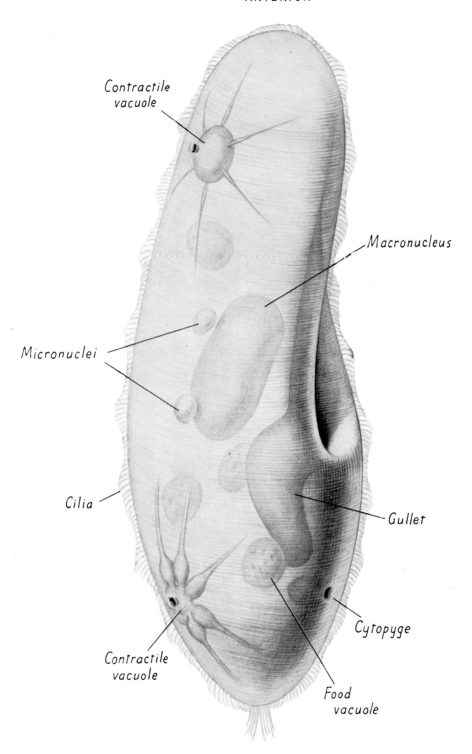

Contractile
vacuole

Macronucleus

Micronuclei

Cilia

Gullet

Contractile
vacuole

Cytopyge

Food
vacuole

POSTERIOR

PLATE 2　The arrangement of the kineties at the surface of *Paramecium* as shown by silver staining

The silver-staining method was that of Klein (1926) carried out on whole-mounted specimens of *P. aurelia* which were flattened on to the microscope slide so that a considerable area was in focus when the photographs were taken by light microscopy. The silver has been deposited mainly along the line of the kineties.

Fig. 1　Light micrograph ×780. The ventral surface as it would appear to an observer below and outside the animal. The anterior pole is toward the top of the page. The opening of the gullet (arrowed) is shaped like a comma and the shorter kineties which run to either side of it, begin and end at the lines of suture. The anterior suture runs along the ventral mid-line between the gullet and the anterior pole, while the posterior suture runs between the gullet and the posterior pole of the paramecium. A line of suture represents a discontinuity between two adjacent fields in each of which the kineties run in parallel rows. At the anterior suture the kineties of the two fields meet at approximately a right angle and at the posterior suture the kineties meet at an angle of approximately 30°. Note that to the paramecium's left side of the gullet (to the right in the picture) the kineties are more curved than to its right.

Fig. 2　Light micrograph ×1800. Part of the ventral surface crossed by the posterior line of suture is shown in the region of the cytopyge, that is midway between the gullet and the posterior pole. The cytopyge is closed and appears here as a line of dense silver deposition (arrowed).

Fig. 3　Light micrograph ×1800. Part of the mid-dorsal surface. The kineties run in parallel rows in the anterior-posterior direction. The dark patch of densely deposited silver (arrowed) probably represents the position of a contractile vacuole.

Fig. 4　Light micrograph ×780. Polar surfaces of two individual paramecia. The kineties meet at a point near the poles which also marks the end of a suture line. In the lower half of the figure the anterior pole of a paramecium is shown. Part of its anterior suture can be seen to run to the pole from the gullet region shown in the extreme lower left-hand corner. The outlines of a number of cytoplasmic crystals can also be seen superimposed on the silver-stain pattern. Crystals can also be seen in Fig. 1 above and Plate 30.

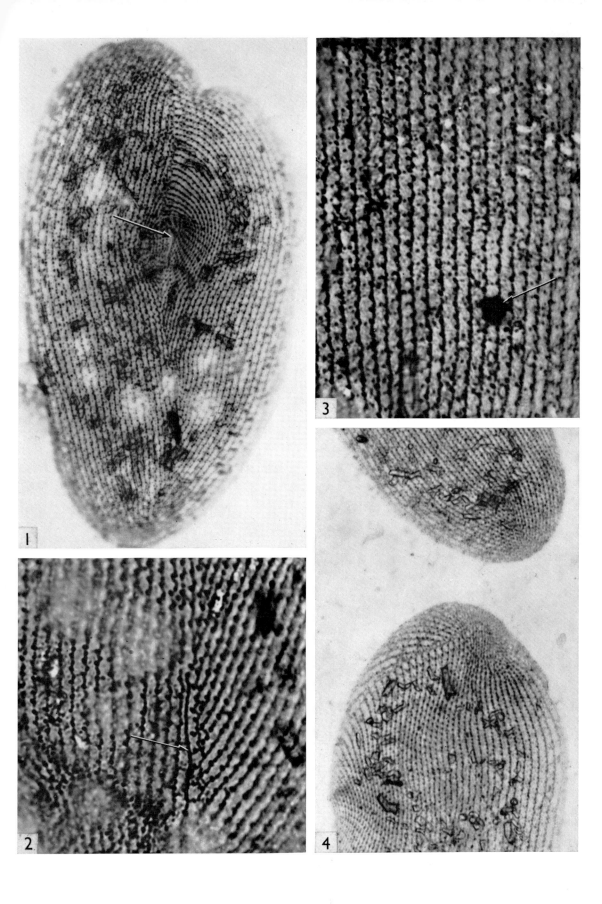

PLATE 3   The pellicle. Pairs of cilia

FIG. 1   Light micrograph × 1500. This photograph is of a stained section cut tangentially to the ventral surface. The anterior pole of the paramecium is to the top of the picture and the section is viewed from outside the animal. The vestibulum (v) of the gullet is in the lower left of the photograph. From the vestibulum the oral groove extends anteriorly and somewhat diagonally. Along the bottom of the groove the plane of the section cuts the ventral surface more superficially, so that the rows of corpuscular units can be seen each with a pair of sectioned cilia (represented by a pair of dots). The many white patches in the cytoplasm are sectioned trichocysts.

FIG. 2   Electron micrograph × 14,000. This section has been cut tangentially to the ventral surface in the region of the anterior suture and is viewed here from inside the paramecium. The anterior end is towards the top of the page. The suture line runs from a point about midway along the top edge of the figure to a point near its bottom right-hand corner. Basal granules are seen in pairs below the cell surface, in the centre of the field. Corpuscular units of the pellicle can be seen with paired cilia (in section) and also there are units with single cilia. Also to be seen are sectioned trichocysts (tt, see also Plate 12), kinetodesmal fibrils (k), parasomal sacs (p) and fragments of endoplasmic reticulum (er).

FIG. 3   Electron micrograph × 3800. This section is cut tangentially to the surface of the paramecium in a region just to the animal's left of the vestibulum (v). The section is viewed here from inside the paramecium and anterior is towards the top of the figure. The cilia and basal granules are all in pairs and the kinetodesmal fibrils can be seen to run anteriorly and slightly to the right from the posteriormost basal granule of each pair.

FIG. 4   Electron micrograph × 30,000. This tangential section is of a region similar to that shown in Fig. 3 but at a greater magnification. In this case, however, the viewpoint is that of an observer outside the paramecium. The corpuscular units of the pellicle each with two cilia are seen to be hexagonally packed, so that each unit is surrounded by six similar ones.

88

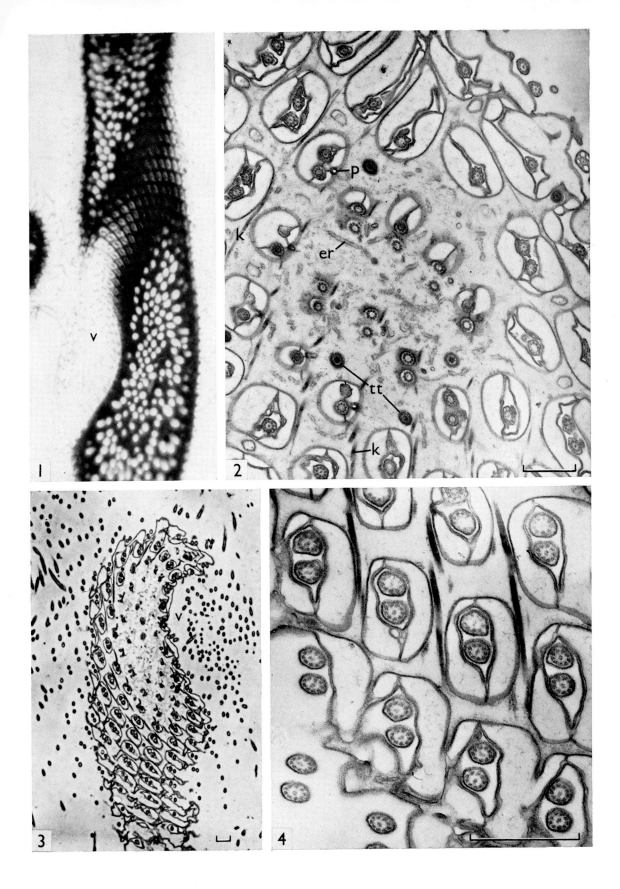

PLATE 4   A general view of the corpuscular units of the pellicle

Electron micrograph ×44,000. This tangential section is viewed from outside the paramecium. The anterior pole of the animal is to the right of the plate. Along the top of the plate the section approaches the surface of the paramecium and here the angular hexagonal border between adjacent corpuscular units can be seen. Interpretation of the infraciliature will be assisted if a comparison is made between this plate and Diagrams 3 and 4. Kinetodesmal fibres (k) run anteriorly from their basal granules (b) to the same side as the parasomal sacs (p). The extreme tops of the sheaths which surround trichocyst tips (tt) are seen here in section situated between the basal granules of adjacent units of a kinety and they appear here as rings of tubules. Transverse tubular fibrils (tf) and postciliary tubular fibrils (pf) are sectioned as they run towards the ridges in which the kinetodesmal fibrils run. Very fine, rather whispy non-tubular fibrils (f) can be seen to radiate in all directions into the cytoplasm from a pair of basal granules somewhat above the midpoint of the plate. There are alveoli (a) to either side of the basal granules.

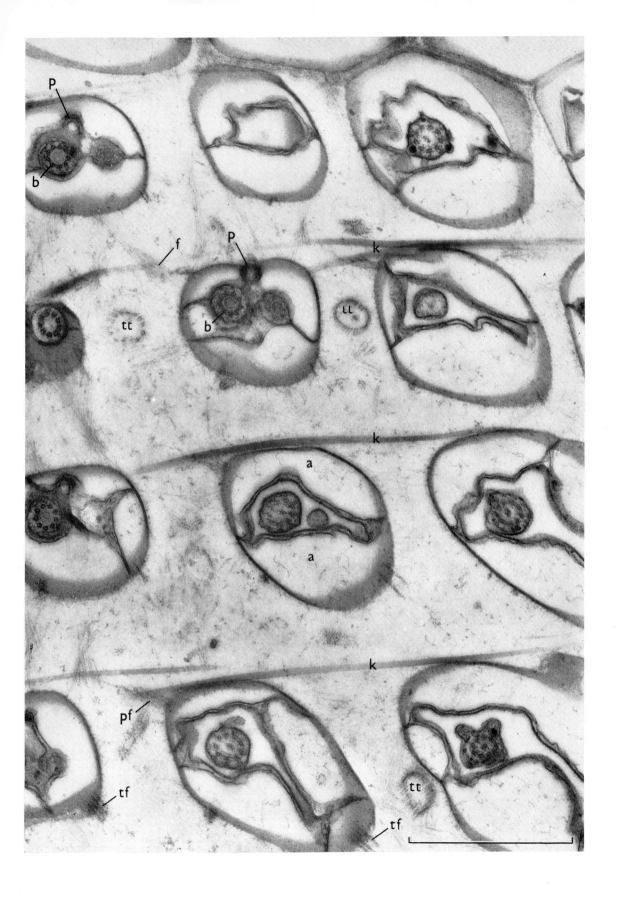

PLATE 5   A tangential section of the cell surface

Electron micrograph × 16,000. This is a composite picture made up of two overlapping electron micrographs. It shows part of the dorsal surface of a paramecium with details from eight adjacent kineties. The top edge of the plate is anterior. Because the surface of a paramecium is rounded and a section is planar, surface material is seen sectioned at different depths over the plate. The plane of the section only passes beneath the bottom of the basal granules over a small area in the upper middle of the plate. Some units of a kinety are seen to be organised about a pair of basal granules while in other cases there is only a single basal granule. Kinetodesmal fibrils may be seen joined to single basal granules or to the more posterior basal granule of a pair. Mitochondria (m) occur at the level of the basal granules or just below. The tips of trichocysts (tt) with their sheaths are seen in section at a lower level than was the case in Plate 4. In this section only about one quarter of the possible trichocyst sites appear to be occupied by a trichocyst. In this section the cytoplasm appears denser than in Plate 4, probably because glycogen has been preserved. However, the better preservation makes it more difficult to see certain details such as fine fibrils, although some bundles of fine fibrils (f) can be seen to run mostly in an anterior–posterior direction at a greater depth than the kinetodesmal fibrils.

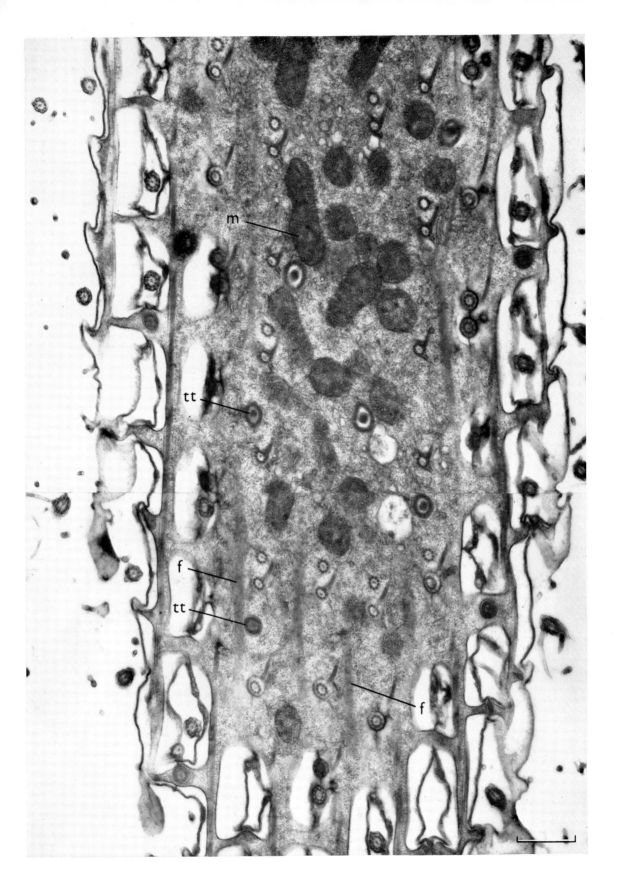

PLATE 6   An oblique section of the cell surface

Electron micrograph × 16,000. This is a composite picture of two adjacent micrographs. The oblique angle of the section gives a greater range of depth over the dorsal cell surface. The section passes through eleven adjacent kineties. The surface structures to be seen are the same as for Plate 5, but in this case the concentration of mitochondria below the level of the basal granules is more obvious. Rough endoplasmic reticulum with ribosomes can be recognised between the mitochondria.

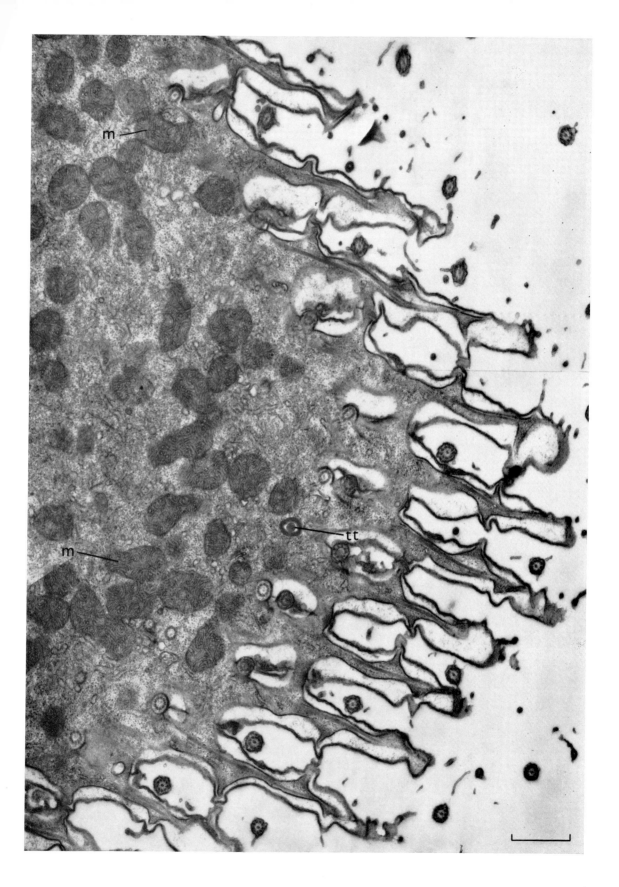

PLATE 7   Cilia and basal granules

These four electron micrographs are of longitudinal median sections through a cilium and its underlying basal granule. The cilia are all situated in the pellicle and a pair of membrane-limited alveoli (a) are to be seen to either side of the proximal part of the cilium where it joins the basal granule. Reference may be made to Diagram 4 to interpret the ultrastructures. The microtubular fibrils, inside the surface of each cilium, can be seen to be continuous with the microtubules which outline the basal granule (b). The plasma membrane at the surface of the pellicle is continuous with the membrane which forms the surface of the cilium.

FIG. 1   Electron micrograph × 32,000. The cilium's axial pair of separated microtubules meet at the base of the cilium at the axial granule (ax), and there is a curved septum across the cilium immediately below the axial granule. The dense fibrogranular layer (d) which runs immediately below the alveoli is continuous with the dense plate which closes the upper surface of the basal granule (see also Figs. 2 to 4). A thinner layer can also be seen immediately above this thicker layer. The membrane of the cilium has a number of small projections which may be artefacts (cf. Pitelka, 1965).

FIG. 2   Electron micrograph × 54,000. Reproduced by courtesy of Mott (1965) to illustrate the ferritin-labelled antibody method of demonstrating the presence of antigenic material on the cell surface of *P. aurelia*. External to the plasma membrane (seen as a pair of fine double lines) are a number of dark dots which represent ferritin molecules (fe), embedded in a fuzz of antibody molecules which have condensed on antigenic sites. The continuity of the plasma membrane (p) and the membrane round the cilium is clear. There are also two transverse sections through neighbouring cilia towards the top of the figure.

FIG. 3   Electron micrograph × 60,000. Structures at the join between the cilium and the basal granule are shown. Below the basal granule is part of a rather abnormally situated trichocyst. To the right of the picture there is a depression in the plasma membrane to a level just below the alveolus and this is almost certainly the opening of the parasomal sac (arrowed).

FIG. 4   Electron micrograph × 36,000. Anterior to the basal granule with the cilium there is a second basal granule which does not appear to have a cilium. In Fig. 1 there is also to be seen part of a second basal granule which has been sectioned at grazing incidence.

96

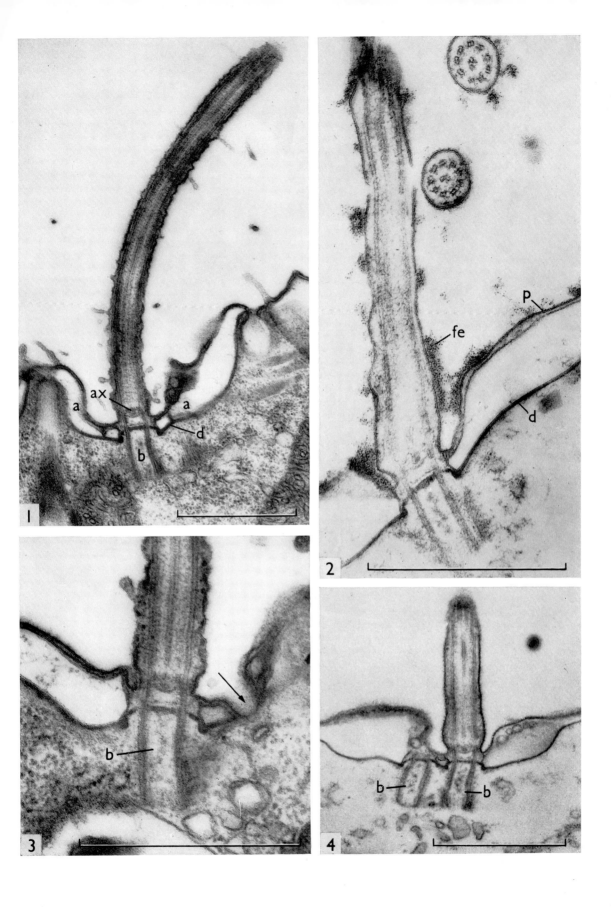

PLATE 8   Unit membranes at the surface of the pellicle

Over most of the cell surface of *Paramecium* there are three membranes. The outermost is the plasma membrane; below this there are the outer and inner alveolar membranes. Immediately below the inner alveolar membrane is a granular layer. After fixation in osmium tetroxide each unit membrane should appear as a three-layer structure under the electron microscope provided that the magnification is sufficiently high and the plane of the section is perpendicular to the plane of the membrane. Each unit membrane consists of two dark osmiophilic layers (2–3 mμ thick) enclosing a light, unstained layer (3 mμ thick).

FIG. 1   Electron micrograph × 130,000. The plasma membrane (pl) and the outer alveolar membrane (oa) both appear clearly as triple-layered unit membranes separated from each other by a very thin layer of cytoplasm about 10 mμ thick. Outside the triple-layered plasma membrane there is a diffuse layer of material which may be the surface antigen. To the left of the field the triple-layered membranes cannot be resolved and they appear diffuse (arrowed) because they have been sectioned obliquely. For the same reason the inner alveolar membrane, which bounds the cytoplasm at the bottom of the field, is also not sharply defined.

FIG. 2   Electron micrograph × 130,000. The inner alveolar membrane (ia) appears as a triple layer and immediately beneath it can be seen a dense osmiophilic fibrogranular layer (d). The plasma membrane and the outer alveolar membrane can also be seen. The latter joins the inner alveolar membrane at the median septum (s), which separates the two alveoli (a) present in each single corpuscular unit of the pellicle. The median septum runs in an anterior-posterior direction down the line of basal granules which is sometimes said to constitute a ciliary meridian. In the bottom right-hand corner is a small part of a trichocyst tip (tt) and its sheath.

FIG. 3   Electron micrograph × 60,000. This is an oblique section which shows the surface membranes and part of a kinetodesmal fibril (k). At a lower level the upper parts of two basal granules (b) are sectioned at the level of the alveoli. The parasomal sac (p) is sectioned at the level of the dense layer where there are ring like thickenings in the wall of the sac and is connected to the basal granule by a very thin layer or bridge of cytoplasm bounded by alveolar membranes. Fine fibrils (f) run in the cytoplasm between the two basal granules.

98

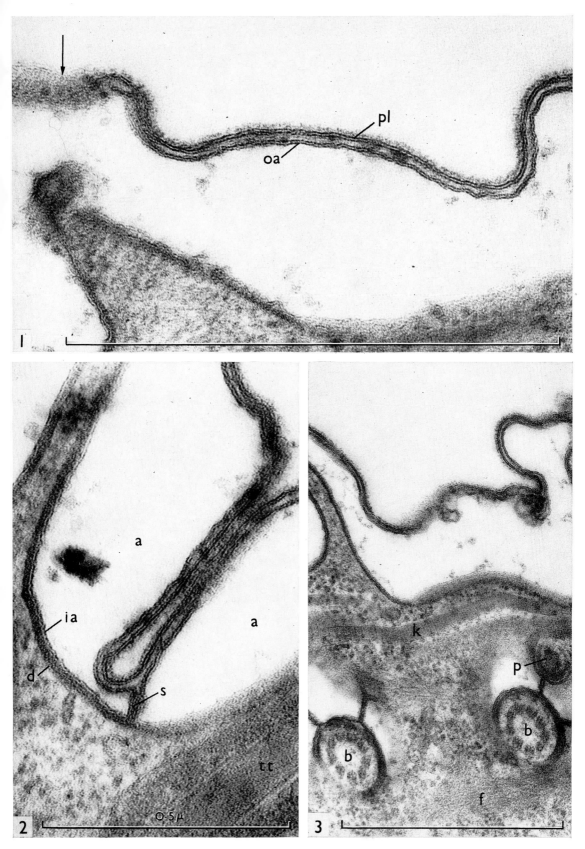

1

pl

oa

2

a

a

ia

d

s

tt

0·5 μ

3

k

p

b

b

f

H

PLATE 9   Kinetodesmal fibrils

FIG. 1   Electron micrograph × 32,000. A section cut transversely through five kineties looking posteriorly. Within the cytoplasmic ridges which divide adjacent kineties, five kinetodesmal fibrils (arrowed in one ridge) may be seen in cross-section in each of three ridges. The kinetodesmal fibrils run one above the other, the lowest having the largest cross-section and the highest having the smallest cross-section. The reason for this is that each fibril tapers and runs nearer the cell surface the further it runs anteriorly from the basal granule of its origin (see Diagram 1). Vesicular cytoplasm can be seen in several places above the median septum in regions bounded by the plasma membrane and the outer alveolar membranes.

FIG. 2   Electron micrograph × 100,000. The figure shows part of a kinetodesmal fibril, sectioned longitudinally to show its regular transverse striations. The major periodicity is at an interval of about 28 mμ and there are darker and lighter sub-bands within this interval. The two smaller kinetodesmal fibrils shown above the main one, also have transverse striations.

FIG. 3   Electron micrograph × 55,000. The figure shows the attachment between a kinetodesmal fibril (k) and the proximal part of a basal granule (b). The longitudinal section is cut transversely to the cell surface. The kinetodesmal fibril is widest where it meets the basal granule. The dense layer which lies just beneath the inner alveolar membrane crosses the distal end of the basal granule. The cilium may have been broken off. In Figs. 2 and 3 the methacrylate embedding medium was used.

FIG. 4   Electron micrograph × 48,000. Transverse section of two kinetodesmal bundles. In this case, which is unusual, the individual kinetodesmal fibrils do not run above each other but are twisted. The median septum (s) can be seen.

FIG. 5   Electron micrograph × 48,000. This transversely sectioned paramecium shows as many as six kinetodesmal fibrils one above the other in the kinetodesmal bundle or kinetodesma.

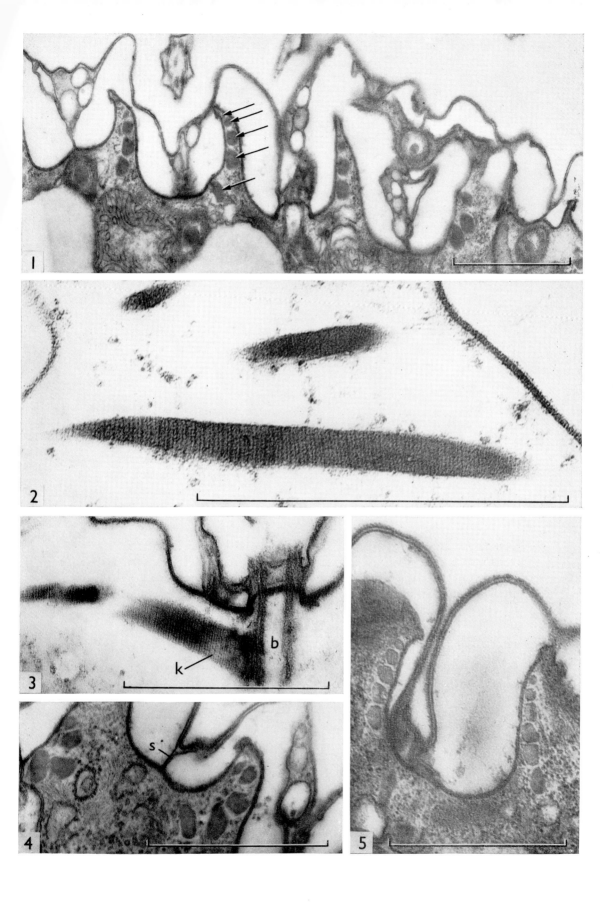

PLATE 10   Pellicle structures including infraciliature

These electron micrographs relate the infraciliature (which comprises the kinetodesmal fibrils, the microtubular fibrils and the fine fibrous network) to the other features of the pellicle such as the basal granules, the alveoli and the longitudinal cytoplasmic ridges.

FIG. 1   Electron micrograph × 36,000. The section shows five units of the pellicle. It is cut almost tangentially to the cell surface, but it is at its greatest depth in the bottom left corner of the picture and the least depth in the top right corner. The section is viewed from outside the paramecium. The anterior of the animal is towards the top right corner of the figure. The kinetodesmal fibrils (k) which may be recognised by their cross striations, run longitudinally in cytoplasmic ridges between the rows of cilia. Four of the units of the pellicle can be seen to have a pair of cilia while a fifth unit has but a single cilium. The cilia are each sectioned at a slightly different level near the basal granule. The pair in the bottom left corner are seen surrounded by the pair of alveoli (a) separated by the median septum (s), whereas the other cilia are immediately surrounded by a boat-shaped space which is outside the plasma membrane. The alveoli surround this depression on the other side of a pair of membranes. Bundles of fine fibrils (f) may be seen near the bottom left corner of the picture and transverse (tf) and postciliary tubular fibrils (pf) are also sectioned obliquely but are difficult to distinguish at three points round the pair of basal granules.

FIG. 2   Electron micrograph × 28,000. The tangential section is cut at the level of the basal granules and is viewed from inside the paramecium. The anterior direction is to the top of the picture. Kinetodesmal fibrils (k) can be seen to run anteriorly from the basal granules to the posteriormost basal granule of a pair. Dense bundles of fine fibrils (f) run mainly between the rows of basal granules and mainly but not entirely in the anterior–posterior direction.

FIG. 3   Electron micrograph × 60,000. This section of one of the cytoplasmic ridges, which run longitudinally between the rows of cilia, shows five kinetodesmal fibrils in cross-section. The fibrils are stacked one above the other with the one with the largest cross-section (k) at the bottom of the ridge and the smallest at the top. The ridge is separated from the alveolus (a) by the alveolar membrane and a dense layer. Immediately to the right of the second largest section through a kinetodesmal fibril, a ribbon of four microtubules can be seen in section; these are postciliary tubular fibrils (pf). Below the level of the kinetodesmal fibrils a bundle of fine fibrils (f) runs transversely.

102

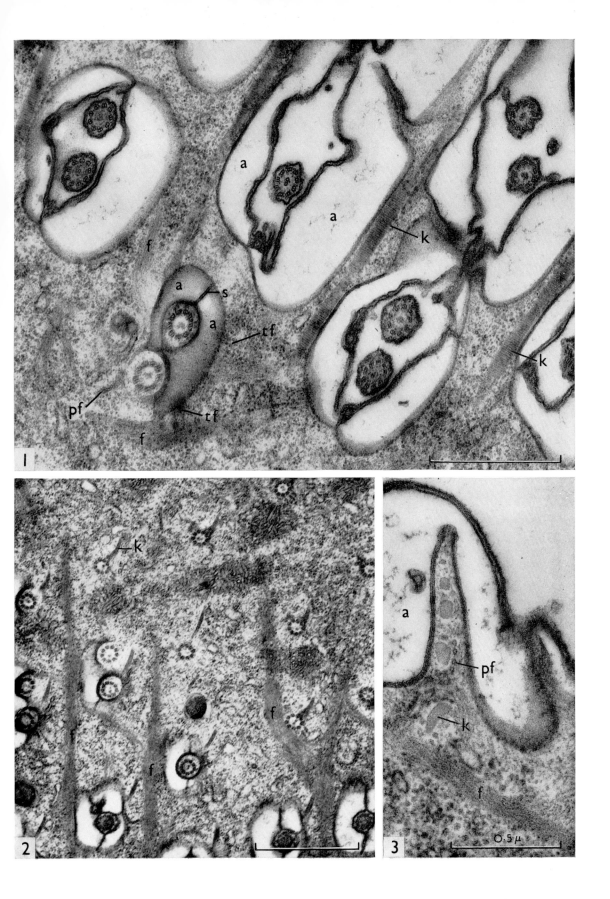

PLATE 11   Sections through basal granules and cilia

Fig. 1   Electron micrograph ×24,000. This transverse section through the pellicle shows two cilia and their basal granules in adjacent kineties.

Fig. 2   Electron micrograph ×50,000. A cilium has been sectioned transversely at its proximal end. The membranes may be interpreted by reference to Fig. 1, as the two figures are of sections cut at right angles to each other.

Fig. 3   Electron micrograph ×50,000. Cilia have been sectioned at different levels. The cilium to the left of the picture shows the ring of nine paired peripheral microtubules and the two separate axial tubules. Such a section is typical of a cilium over most of its length. The other cilia are sectioned near their tips where the number of tubules is reduced. The peripheral tubules have become single instead of paired.

Fig. 4   Light micrograph ×1200. This phase contrast picture is of the posterior end of a live paramecium, immobilised by slight compression between a slide and a coverslip. The long cilia at the posterior end can be seen because they were relatively immotile over the photographic exposure time of several seconds whereas the motile cilia over the remainder of the surface cannot be seen because they were moving.

Fig. 5   Electron micrograph ×50,000. A tangential section through the lower half of a pair of basal granules shows the kinetodesmal fibril (k) attached to the posteriormost basal granule. Linking the basal granules, on the side opposite the kinetodesmal fibril, can be seen a bridge (br) of fibrillar material.

Fig. 6   Electron micrograph ×27,000. A tangential section through the dorsal surface of a paramecium. The opening to the right is the discharge channel of a contractile vacuole.

Fig. 7   Electron micrograph ×50,000. A section cut parallel to the surface through basal granules in adjacent kineties. Transverse tubular fibrils (tf) and posterior tubular fibrils (pf) originate at the proximal part of a basal granule. Nine triplets of microtubules are arranged at the periphery of a basal granule (b) and they resemble a water-wheel.

Fig. 8   Electron micrograph ×50,000. A transverse section through the base of a pair of cilia at the level of the alveoli (a) to show the membrane limited cytoplasmic connection between the posterior basal granule (b) and the parasomal sac (p).

Fig. 9   Electron micrograph ×65,000. A section cut through two pairs of basal granules, and viewed from the interior of the cell. Kinetodesmal fibrils (k) attach to the most posterior basal granule of each pair. Posterior and transverse tubular fibrils can be seen.

Fig. 10   Electron micrograph ×32,000. A pair of cilia in section at their proximal end at the level of the alveolus showing the connection with the parasomal sac (p).

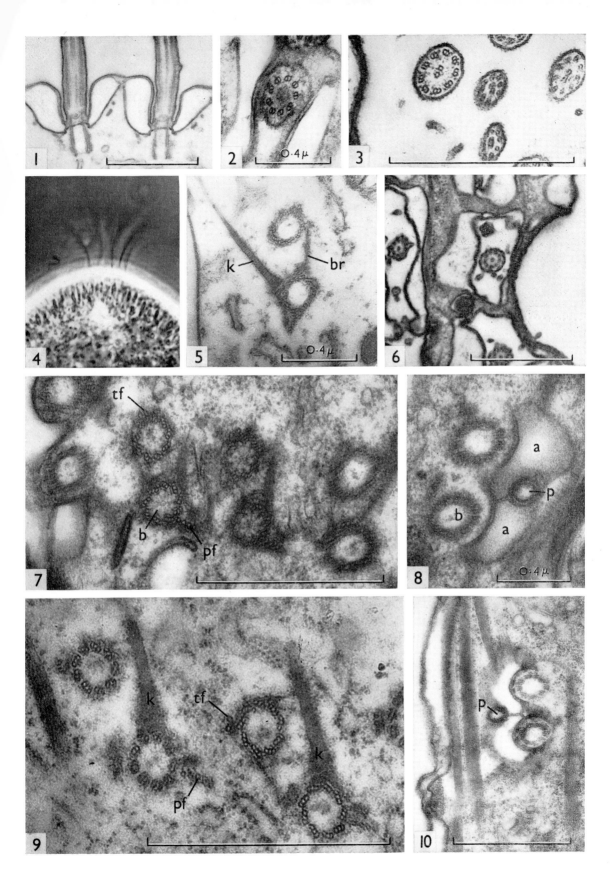

PLATE 12   Mature trichocysts

Fig. 1   Electron micrograph ×25,000. A longitudinal section of a mature trichocyst (mt) in its position below the pellicle. The white appearance of the carrot-shaped body should not be taken as indicative of lack of substance but demonstrates a lack of affinity for osmium tetroxide. The trichocyst tip and its sheath are darkly stained with the osmic fixative.

Fig. 2   Electron micrograph ×11,000. A tangential section through the pellicle to show the positions down the rows of cilia which are occupied by the tips of mature trichocysts (tt). The trichocysts are seldom displaced far from a line drawn through the basal granules of a kinety.

Fig. 3   Electron micrograph ×90,000. A longitudinal section through the proximal part of a trichocyst tip (tt). The herring-bone pattern indicates that its constituent macromolecules are ordered in a crystalline lattice. There appears to be a membrane along the surface of the trichocyst tip.

Fig. 4   Electron micrograph ×60,000. A longitudinal section through the osmiophilic trichocyst tip (tt) and its sheath in its position below the pellicle. The sheath has an irregularly granular ultrastructure. Part of the apparently structureless body of this mature trichocyst appears greyish white at the bottom of the picture. The trichocyst is between two rows of kinetodesmal fibrils (k).

Fig. 5   Electron micrograph ×60,000. This section is transverse to a trichocyst tip and at the level of the alveoli (a) of the pellicle. The section is not quite parallel to the cell surface. The trichocyst tip is surrounded by its sheath which shows a ribbed outer surface to the surrounding cytoplasm. The mature trichocyst always meets the cell surface between alveoli, as was also shown in Figs. 1, 2 and 4.

Fig. 6   Electron micrograph ×60,000. This transverse section through a trichocyst tip and its sheath is cut at a level further from the cell surface than is Fig. 5 and shows a space between the tip and the sheath. Both the tip and the sheath have an outer membrane.

106

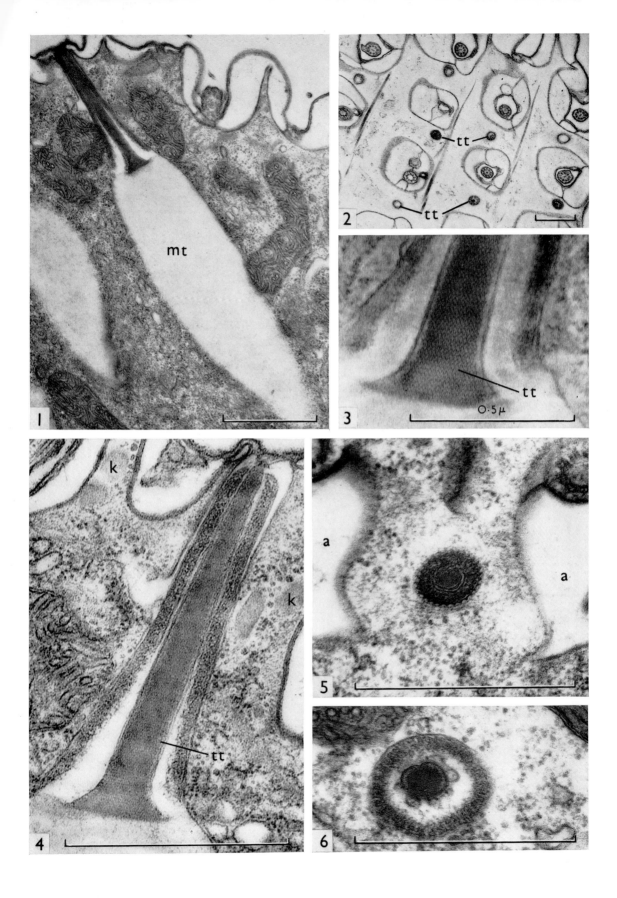

PLATE 13   Extruded trichocysts

Mature trichocysts may be extruded into the medium when stimulated by acid or by electrical impulses and they may also be extruded backwards into the cytoplasm of the paramecium by the action of certain cold fixatives which penetrate slowly. On extrusion, the body of the trichocyst is changed to a shaft and when fully extended the shaft shows regular cross-striation illustrated by the electron micrographs of this plate. The trichocysts of Plate 12 (except for Figs. 2 and 3) were fixed in warm osmic fixative at 37° C to obviate extrusion.

FIG. 1   Electron micrograph ×26,000. The trichocyst is shown in its usual position under the pellicle. The body of the trichocyst has been partly but not fully extruded backwards by the action of the fixative and shows some weak and somewhat irregular cross striation.

FIG. 2   Electron micrograph ×15,000. The trichocyst is partly extruded backwards as in Fig. 1. The body is longer and thinner than normal and has a pronounced curvature. Some cross striations are visible.

FIG. 3   Electron micrograph ×24,000. This trichocyst is shown in longitudinal section with its shaft fully extended backwards into the cytoplasm. Regular cross-striations can be seen with a periodicity of 60 mμ. The structure of the trichocyst tip and its sheath, however, remains unchanged. The cytoplasm surrounding the trichocyst has been incompletely preserved by the methacrylate embedding procedure employed.

FIG. 4   Electron micrograph ×70,000. A longitudinal section through part of the shaft of a fully extruded trichocyst. The regular cross-striations show a periodicity of 50 mμ. Longitudinal fibres also appear to be present in the shaft and these are about 3 mμ in diameter.

FIG. 5   Electron micrograph ×190,000. A negatively stained tip of an extruded trichocyst, showing regular cross-striations at intervals of 16 mμ and a herring-bone pattern of structure similar to that shown in Plate 12, fig. 3. This electron micrograph is reproduced by courtesy of Dr T. F. Anderson and Dr J. R. Preer. Cultures of *P. aurelia* were lysed in sodium deoxycholate and homogenised; the specimens were prepared according to the method of Anderson, Preer and Bray (1964).

FIG. 6   Electron micrograph ×120,000. A small part of an extruded trichocyst shaft which shows cross-banding with a main periodicity of 60 mμ, although there are many minor cross bands to be seen within the main interval. The method of preparation was the same as for Fig. 5. The micrograph is reproduced by courtesy of Dr T. F. Anderson and Dr J. R. Preer.

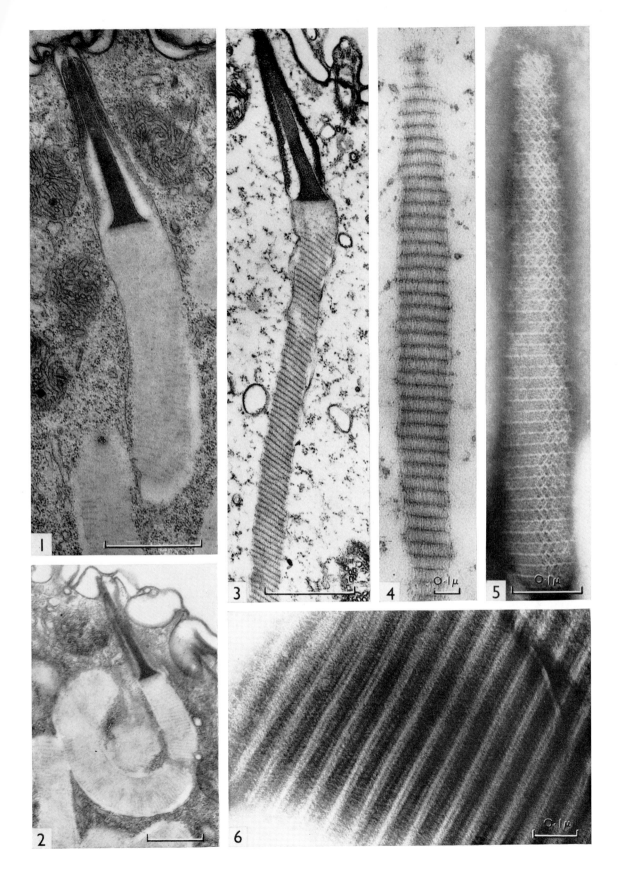

1

2

3

4   0·1μ

5   0·1μ

6   0·1μ

PLATE 14   Gullet and vestibulum in cross-sectioned

Paramecium

Fig. 1   Electron micrograph × 5000. A nearly equatorial cross-section through a specimen of *P. aurelia*. The section includes the gullet opening where the gullet joins the
flared vestibulum (v) which is part of the pellicle. The section is through the widest
part of the paramecium and the position of the section can be gauged by comparison
with Plate 1. A considerable area in the middle of the section is occupied by the macronucleus (ma) with its large and small bodies (see Plate 33). A micronucleus (mi) is below
and almost in contact with the macronucleus. A few irregular white holes (cr) represent
the position of crystalline inclusions (see Plate 30). The more numerous and regularly
rounded white bodies below the pellicle are mature trichocysts. Mitochondria may be
recognised as numerous small dark bodies just below the pellicle. Such relatively low-
power electron micrographs allow an estimate to be made of the total number of
kineties in the whole paramecium. There are about eighty-eight kineties in this case.

Fig. 2   Electron micrograph × 4200. This cross-section through a paramecium has been
cut slightly nearer the posterior end than was the section in Fig. 1. The gullet opening
into the vestibulum is therefore not shown, and the gullet appears as a rounded opening
surrounded by cytoplasm. On the innermost wall of the gullet are rows of cilia which
constitute the peniculus and quadrulus (Plates 15 to 18). Note the macronucleus. The
considerable number of mature trichocysts (white bodies below the pellicle) indicates
that a large part of the cell's metabolism must be directed towards the synthesis of these
bodies.

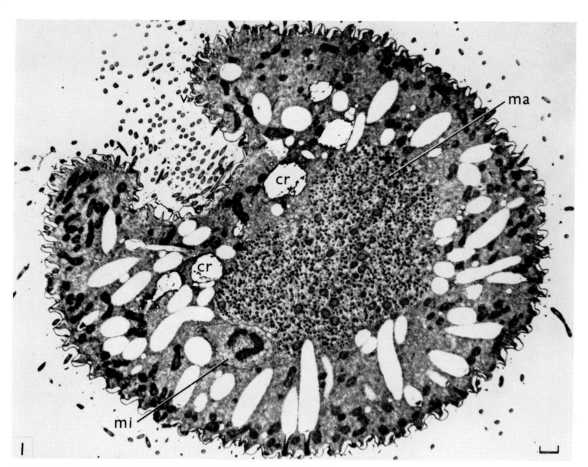

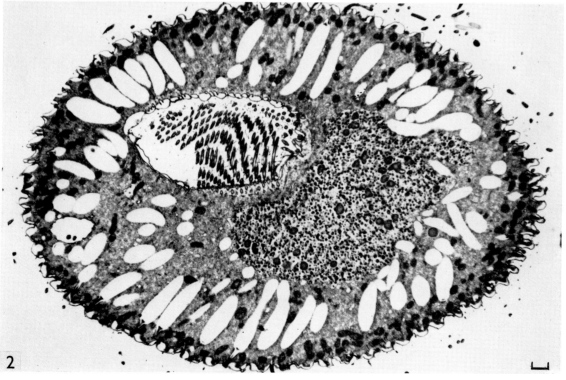

PLATE 15   General morphology of the gullet

FIG. 1   Light micrograph × 1500. A longitudinal section through a paramecium fixed and embedded as for electron microscopy but sectioned at about 1 μ thickness and stained with toluidine blue. The gullet opening is in the ventral surface of the paramecium. The rows of cilia which make up the peniculus and quadrulus can just be distinguished as they run down the inner wall to the bottom of the gullet where food vacuoles form. Food vacuoles can be seen in the cytoplasm as well as the macronucleus and a micronucleus. A majority of trichocysts are darkly stained and therefore juvenile (that is, immature, see Plates 41 to 43) so that this paramecium must have been fixed within an hour before or after fission.

FIG. 2   Light micrograph × 3900. This is a detail from a photograph taken of a whole-mounted paramecium fixed and silver stained as for Plate 2. The viewpoint is outside the ventral surface looking into the gullet and four dark rows of dots represent the four ciliary rows of the quadrulus (q) which is on the inner wall of the gullet opposite the opening.

FIG. 3   Electron micrograph × 3000. This electron micrograph is from a section cut at a similar orientation to Fig. 2. Eight close-packed rows of cilia and their basal granules which form the peniculus (pn) are clearly shown and the sectioned cilia form a ribbon across the picture. The four ciliary rows of the quadrulus (q) are more widely spaced in this anterior half of the gullet. The ribbed wall (r) forms a gullet surface at the top of the figure and a row of alveoli (a) separates the cytoplasm from the gullet in this region. The surface at the bottom of the figure is part of the vestibular region of the pellicle.

FIG. 4   Electron micrograph × 10,000. This shows part of an isolated gullet which was found as a contaminant in a pellet of isolated macronuclei prepared by Stevenson (1967b). The gullet has completely separated from the pellicle, but the wall of the gullet has held its structures together. Recognisable features include the ribbed wall (r), the rows of basal granules from the peniculus (pn) and quadrulus (q) and the naked ribbed wall (n). Parts of two isolated macronuclei are on either side of the isolated gullet.

112

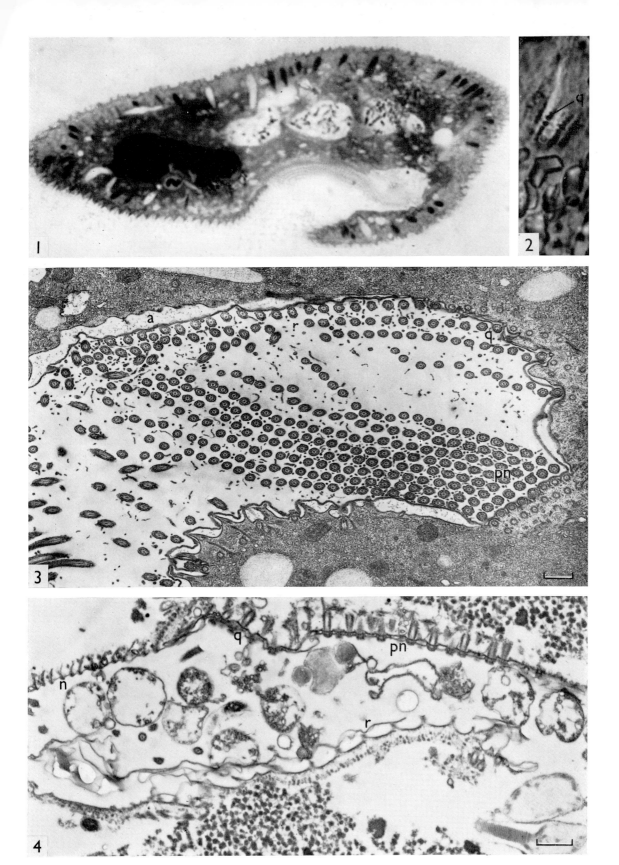

PLATE 16 Peniculus, quadrulus, ribbed wall and endoral membrane

Fɪɢ. 1 Electron micrograph × 12,000. This section is transverse to the paramecium at the level of the gullet opening. The vestibular region of the pellicle is to the top of the picture. The various parts of the gullet are seen from the point of view of an observer looking posteriorly. On the left of the figure, which is also the left side of the paramecium, the closely packed ciliary rows of the peniculus (pn) may be seen. On the dorsal side of the gullet at the bottom of the figure and opposite the gullet opening are the four rows of cilia which represent the quadrulus (q). Further round the gullet and opposite the peniculus is the ribbed wall (r) with its alveoli and ridges of cytoplasm. Between the ribbed wall and the vestibular surface, along the right side of the gullet, a double row of cilia runs between cytoplasmic ridges. This is the endoral membrane (e) and a few basal granules from the endoral membrane can be seen in this section. The trichocysts (mt) as in the right of the picture do not position themselves below any part of the gullet or vestibulum but are associated with the ventral parts of the pellicle or occasionally they are found in the endoplasm. The macronucleus is in the left bottom corner of the figure.

Fɪɢ. 2 Electron micrograph × 60,000. This is an enlargement of part of the gullet surface similar to that just to the right of the quadrulus in Fig. 1 at the border of the ribbed wall. The cytoplasmic surface shows inpocketing between ridges but the pockets are part of the alveolar space which is separated from the gullet cavity by two membranes seen in transverse section in the top right of the picture. The surface of the ribbed wall therefore resembles that of the pellicle except that basal granules and kinetodesmal fibrils are absent here.

Fɪɢ. 3 Electron micrograph × 40,000. This detail shows a region similar to that of Fig. 2 but the section is viewed looking anteriorly. Two small invaginated pouches appear to connect with the gullet cavity but in fact the greyish area at the mouth of the pouches probably represents alveolar and plasma membranes which have been obliquely sectioned.

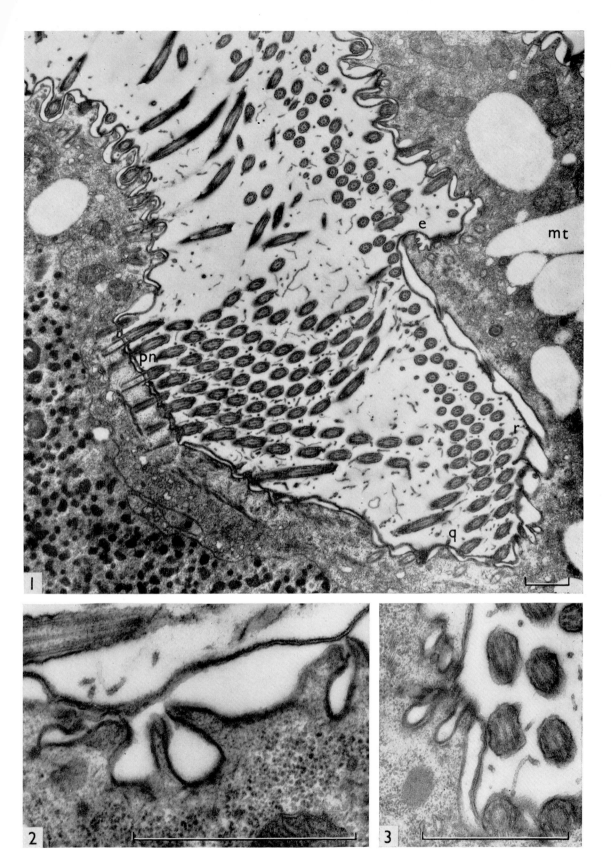

PLATE 17   The gullet and underlying fibrillar networks

Fig. 1   Electron micrograph × 8000. This section is cut at a similar orientation to the section in Plate 16, fig. 1 but here the section is viewed looking anteriorly. The peniculus (pn) has a band of eight rows of hexagonally packed cilia. The upper four rows in the picture constitute the ventral peniculus and the lower four rows make up the dorsal peniculus. To the bottom of the figure is the quadrulus (q). The ribbed wall (r) is on the opposite side of the gullet to the basal granules of the peniculus. Between the ribbed wall and the vestibular part of the pellicle a cytoplasmic promontory projects into the gullet and immediately above it a double row of cilia and basal granules can be seen which form part of the endoral membrane (e). Adjacent to the ribbed wall is the site (arrowed) where the new gullet will develop before the next fission. In the cytoplasm below the ribbed wall run fibrillar networks (f) which are shown at a higher magnification in Figs. 2 and 3.

Fig. 2   Electron micrograph × 27,000. This is an enlargement of an area similar to that below the ribbed wall in Fig. 1. Part of the ribbed wall is shown at the top of the figure, and three cytoplasmic projections are shown between alveoli. The fibrillar networks are shown at two levels separated by part of a trichocyst body (mt).

Fig. 3   Electron micrograph × 40,000. A fibrillar network beneath the ribbed wall of the gullet shows the regular pattern of dark knotting which occurs where the fibres cross. Between double rows of larger knots there are three rows of smaller knots. The whole pattern repeats itself every 400 mμ (see also Diagram 7). Similar patterned fibrillar networks were observed by Schneider (1964b) beneath the basal granules of the peniculus as well as below the ribbed wall. To see the regularity of the pattern in the fibrillar networks it is necessary to cut the section in the correct orientation, otherwise the fibrils appear disordered. Such apparently disordered fibrillar networks appear below the peniculus and quadrulus in Plate 16, fig. 1.

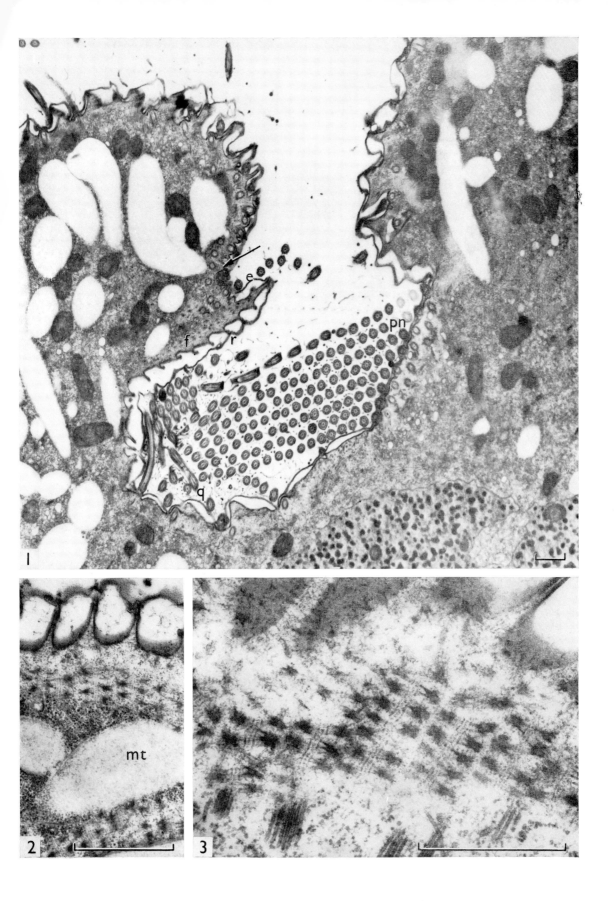

PLATE 18 Basal granules and infraciliature below the peniculus

FIG. 1 Electron micrograph × 10,000. The section is cut through the basal granules beneath a considerable length of the peniculus as it runs down the dorsal wall of the gullet. There are eight parallel rows of basal granules. The section is nearest to the gullet cavity in the middle of the figure which is cut at the level of the alveolus, where the basal granules join the cilia. The rather greyish background to about ninety basal granules sectioned at their distal ends (in the picture's central region), is caused by the septa across the tops of the basal granules and the alveolar membranes which also lie in the plane of the section. The basal granules to either side of the central region are sectioned at a lower level which is further from the surface of the gullet; at the extreme top left- and bottom right-hand corners are fibrillar networks (f) which run below the level of the basal granules. Another fibrillar network runs between the basal granules at a level just below the inner alveolar membrane. Root fibrils (rf) which are always directed towards the quadrulus run for about 500 mμ from the foot of the basal granules in the fourth and eighth rows of cilia. In this picture they project towards the macronucleus from the nearest row of basal granules. Rows of parasomal sacs (p) are also to be seen on the same side of the fourth and eighth rows of basal granules as the root fibrils.

FIG. 2 Electron micrograph × 24,000. This detail shows the line of root fibrils extending from the basal granules of the eighth row of the peniculus. Only the basal granules of the fourth and eighth rows have root fibrils. The alveoli just below the gullet surface and the fibrils below the alveoli are obliquely sectioned.

FIG. 3 Electron micrograph × 32,000. This section has been cut at an orientation similar to that of Fig. 1 only rather more obliquely to the gullet surface at the peniculus; it shows ultrastructural details immediately below the level of the basal granules. The root fibrils (rf) are composed of a single layer of short microtubules arranged transversely or obliquely to the length of each root fibril. Fibrillar networks (f) are represented by grey patches immediately below the basal granules but show no pattern as they are sectioned obliquely. Note the differences in the cross-sections through the basal granules and the proximal region of the cilia at different levels.

118

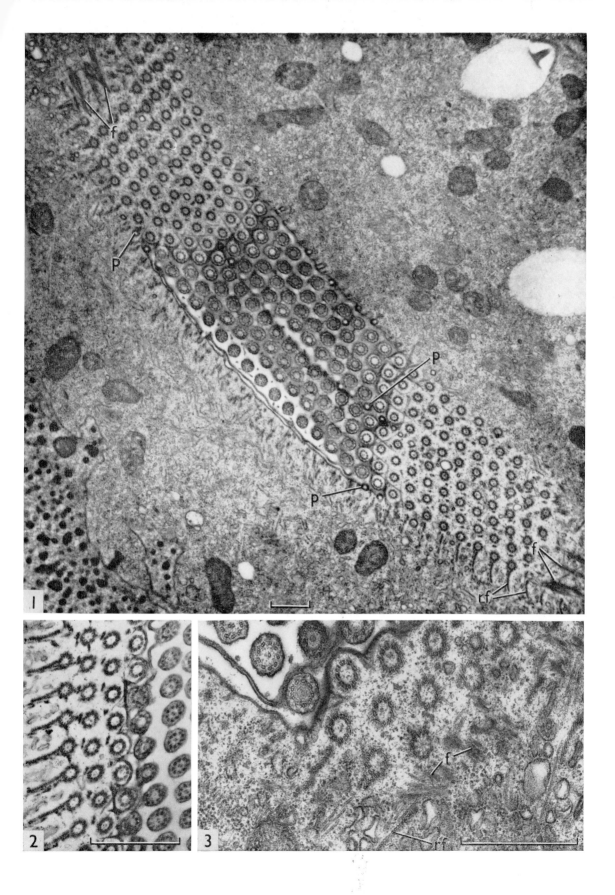

PLATE 19 Microtubules of the postesophageal fibres

FIG. 1 Electron micrograph ×5600. This relatively low power electron micrograph of a longitudinal section shows the postesophageal fibres (pe) of Lund (1933) which run beneath the ribbed wall of the gullet at the left of the figure and extend posteriorly into the cytoplasm beyond the gullet, to the right. The postesophageal fibres were observed by Lund using light microscopy.

FIG. 2 Electron micrograph ×34,000. This is part of the same section shown in Fig. 1 but at a higher magnification. The postesophageal fibres are seen to consist of bundles of parallel microtubules. One microtubular bundle might be equivalent to one fibre observed by light microscopy.

FIG. 3 Electron micrograph ×40,000. This section has been cut transversely to the postesophageal fibres, so that the individual microtubules appear in cross-section as hollow rings 25 mµ in diameter. The microtubules are closely packed in bundles of about 25. Within a bundle the microtubules are packed hexagonally with a centre to centre distance between 35 mµ and 40 mµ.

FIG. 4 Electron micrograph ×20,000. The microtubular bundles are here sectioned obliquely as they run below the surface of the gullet.

FIG. 5 Electron micrograph ×40,000. Some microtubules from the postesophageal fibres run below the ribbed wall region of the gullet. An isolated disc shaped body (see Plate 20) can also be seen in the cytoplasm near the bottom right-hand corner and a few more disc-shaped bodies can also be seen on the left side of Fig. 4.

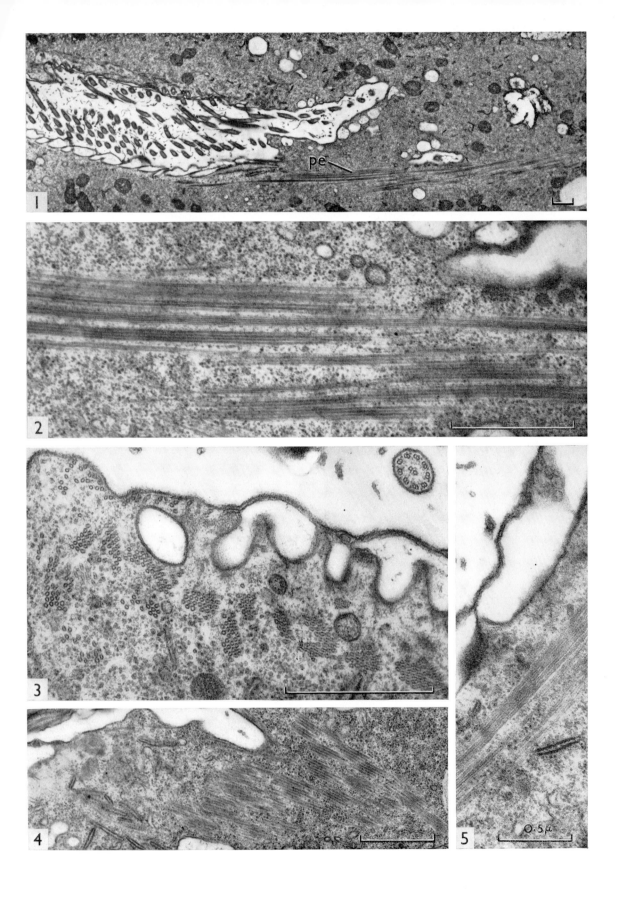

PLATE 20   The naked ribbed wall

At the posterior end of the gullet there is a roughly triangular area of gullet surface called the naked ribbed wall which unlike the rest of the gullet wall is not covered by alveoli. Its apex is about half way up the dorsal surface of the gullet and it borders with the quadrulus and the ribbed wall. Food vacuoles are formed below the naked ribbed wall.

FIG. 1   Electron micrograph × 40,000. The naked ribbed wall forms the gullet surface along the top of the figure. Sheets of parallel microtubules extend beneath the ridges in the surface and are here shown in cross-section as lines of small circles below each ridge. Many disc-shaped bodies (di) show an elongated profile in cross-section, because they are oriented in parallel with the microtubular sheets between which they are found. To the right of the picture, a part of a cord (co) is shown, about 250 mμ wide, but it shows no definite ultrastructure in this case.

FIG. 2   Electron micrograph × 20,000. At the bottom of the picture is the naked ribbed wall with ultrastructures as described for Fig. 1. To the top of the picture, part of the ribbed wall is shown with its superficial alveoli (a) and postesophageal fibres (pe). The two surfaces meet at the left of the picture, and the transversely sectioned cilia which can be seen in the gullet in both Fig. 1 and Fig. 2 are from the nearby quadrulus.

FIG. 3   Electron micrograph × 80,000. This section was cut at right angles to Fig. 1 and shows, in the plane of the section, the microtubular sheets (ms) and also four disc shaped bodies (di) which here appear as rough circles.

FIG. 4   Electron micrograph × 45,000. This section was cut at right angles to the plane both of Fig. 1 and Fig. 2 and just below and parallel to the naked ribbed wall. The disc-shaped bodies are lined up along and parallel to a transversely sectioned sheet of microtubules.

FIG. 5   Electron micrograph × 29,000. This section is cut parallel to Fig. 1 and shows a similar region below the surface of the naked ribbed wall. The background cytoplasm has, however, been partly extracted by a methacrylate embedding procedure and microtubular sheets are seen against a whiter background. Also the cord (co) appears to be composed of fine fibrils which were not apparent in the structure of the cord in Fig. 1.

FIG. 6   Electron micrograph × 80,000. This detail shows a transversely sectioned disc shaped body. It is clearly bounded by a unit membrane within which there is a dark dense layer; there is also a central region of lesser density. A thin dark lamina appears to connect adjacent microtubules in the sheets of microtubules.

122

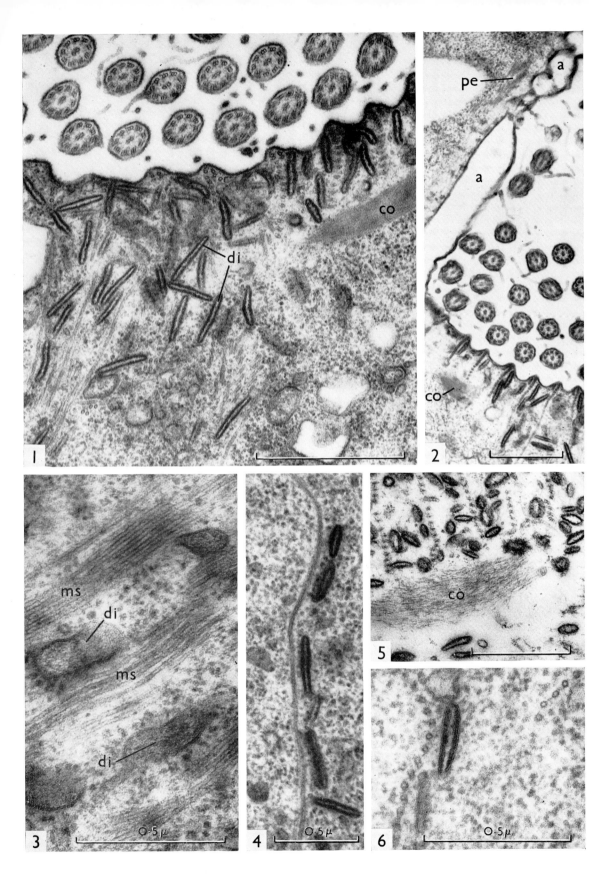

PLATE 21   Fine fibrils below the naked ribbed wall

Fig. 1   Electron micrograph × 20,000. Many disc-shaped bodies (di) are shown below the naked ribbed wall and between the sheets of microtubules. The cytoplasm has been partially extracted and, as a result, the shape of the disc shaped bodies is easier to see, especially when the disc lies in the plane of the section. There is a fibrillar cord (co) in the centre of the field and three small vacuoles in the cytoplasm along the lower border of the picture.

Fig. 2   Electron micrograph × 24,000. This section shows a similar field to Fig. 1 but the standard method of embedding in Araldite has been employed and this gives superior preservation although certain structures visible in Fig. 1 are not apparent here.

Fig. 3   Electron micrograph × 29,000. This section through the cytoplasm immediately below the naked ribbed wall is from partially extracted material and has been selected to show the network of fine fibrils and their connections to the microtubular sheets as a line of dense knotted regions. Between adjacent microtubular sheets and their dark fibrillar knots a triple row of much smaller knots or interconnections between fibrils may be seen. These triple lines are similar to those in the fibrillar networks below the ribbed wall in Plate 17.

Fig. 4   Electron micrograph × 30,000. This section shows a similar field to Fig. 3, but in this case the embedding was done in Araldite which gave superior preservation of cytoplasm. The association between fine fibrils and sheets of microtubules can be seen.

Fig. 5   Electron micrograph × 21,000. This section shows cytoplasm just below the quadrulus of the gullet. It shows a complicated array of fine fibrils but there is no obvious pattern about their arrangement, probably because the section has not been cut in the correct orientation to reveal it.

124

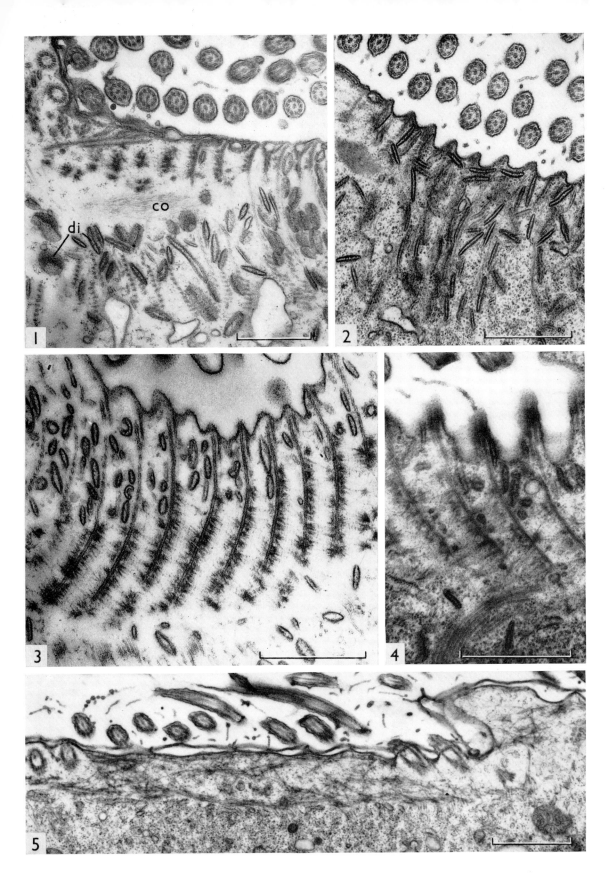

PLATE 22   The formation of food vacuoles

FIG. 1   Light micrograph ×1500. A longitudinal section stained with toluidine blue shows the gullet (g) connected with the ventral surface of the paramecium. At the posterior end of the gullet a food vacuole may be seen in process of formation. The dark bodies in the gullet and in the future food vacuole are bacteria on which paramecium feeds.

FIG. 2   Electron micrograph ×5000. A relatively low-power micrograph of an oblique section to show food vacuole formation at the posterior end of the gullet. This electron micrograph may be compared with the light micrograph, Fig. 1. A food vacuole with bacteria is shown in formation and it lies below the naked ribbed wall (n) and is connected with the posterior end of the gullet (g) by a passage way called the cytopharyngeal tube down which cilia appear to propel the bacteria. Two other food vacuoles (fv) in the cytoplasm are fully formed and enclosed. The anterior end of the gullet is in fact fully open at the ventral surface of the paramecium and only appears closed in this picture because of the angle at which the section was cut.

FIG. 3   Electron micrograph ×20,000. This shows the region of the cytopharyngeal tube during food vacuole formation. The posterior end of the gullet lies to the top of the picture and the future food vacuole which contains bacteria is at the bottom. To the left of the tube the cytoplasm contains disc-shaped bodies (di) and microtubules characteristic of the naked ribbed wall, and to the right there is a part of the postesophageal fibres (pe) which extend posteriorly from beneath the ribbed wall.

FIG. 4   Electron micrograph ×16,000. The new food vacuole is shown with the cytopharyngeal tube about to close above it. Disc-shaped bodies can be seen to the left of the tube and microtubular postesophageal fibres to the right. Note that in Figs. 2, 3 and 4, ultrastructures of the naked ribbed wall were observed on one side of the cytopharyngeal tube and the postesophageal fibres on the other side.

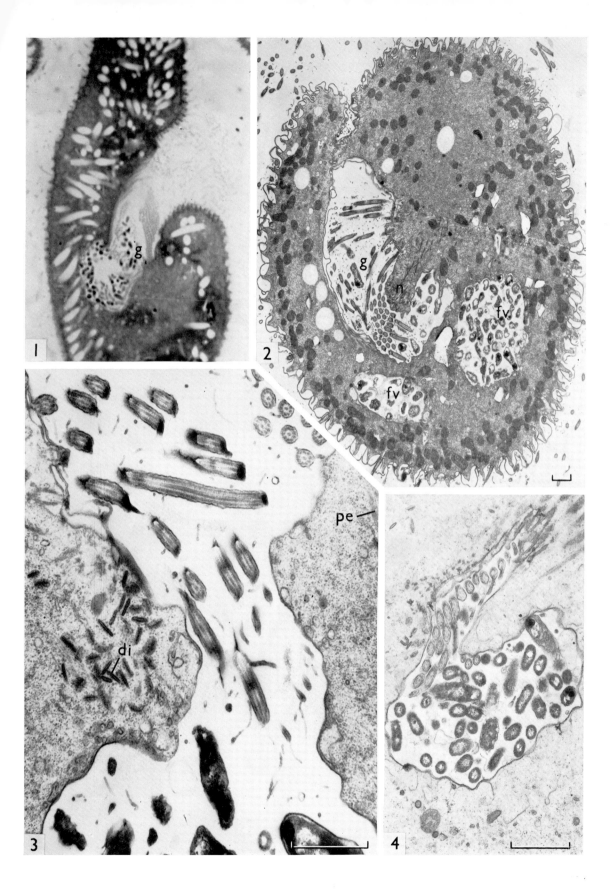

PLATE 23   Food vacuoles in their early stages

FIG. 1   Light micrograph × 2250. Part of a longitudinal section through a paramecium shows two food vacuoles packed with bacteria. In this and in the other figures on this plate the bacteria are mostly intact, which indicates that the process of digestion which takes place in food vacuoles is not well advanced. For this reason the food vacuoles are described as being in their early stages. Part of the gullet (g) is seen to the left of the picture. Toluidine blue was used to stain the microscope slide.

FIG. 2   Electron micrograph × 16,000. This early food vacuole is full of bacteria. Most of these have a normal appearance but some have their membranes raised from the underlying cytoplasm. Also present are some bacterial membranes where the cytoplasm seems to have been lost; there is colloidal material in the vacuole which may have been derived from bacterial cytoplasm. These observations indicate that bacteria are broken down in the food vacuole.

FIG. 3   Electron micrograph × 40,000. Part of the wall of a food vacuole is shown at about the same stage as Fig. 2. The single unit membrane which forms the wall of the food vacuole runs across the picture. In these early stages the food vacuole is quite smooth and not convoluted. Above the membrane some bacteria can be seen in the food vacuole. Immediately below the membrane of the food vacuole can be seen some dense membrane-limited bodies (arrowed) which appear similar to the enzyme granules of Schneider (1964b). It has been suggested that these granules discharge their contents into the food vacuole to provide enzymes to break down the bacteria but there is no direct evidence for this.

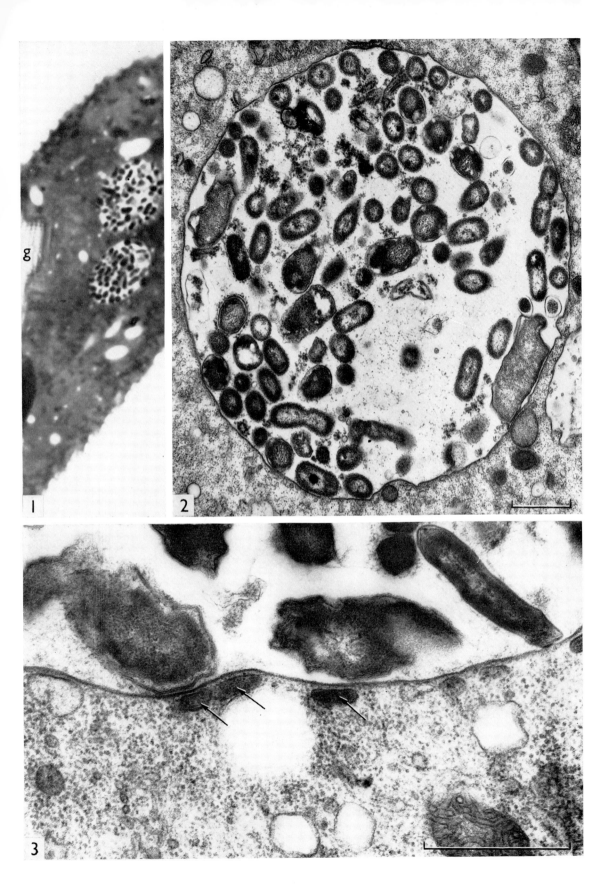

PLATE 24   Food vacuoles at later stages—pinocytosis

Fig. 1   Electron micrograph ×24,000. This shows part of a food vacuole at a rather later stage than those of Plate 23. The bacteria in the vacuole are mostly represented by fragments and the vacuolar membrane is very convoluted in contrast to the earlier stages. The membrane has many flask-like protuberances into the cytoplasm and small secondary vacuoles can be seen in the cytoplasm which were probably formed from the flask-like protuberances by pinching off at the neck. This process is often termed pinocytosis.

Fig. 2   Electron micrograph ×90,000. This is an enlargement of part of the field of Fig. 1. Two secondary vacuoles are bounded by a unit membrane which resembles the membrane of the main food vacuole. Two membrane limited bodies are present and these may be enzyme granules (arrowed).

Fig. 3   Electron micrograph ×60,000. Two flask shaped protuberances are shown which have formed in the membrane of the main food vacuole at the top of the picture. Similar fragments of colloidal material are found in the main vacuole as in the protuberances. The secondary vacuole to the left already appears to be collapsing because one side of it is folded in towards the opposite side.

Fig. 4   Electron micrograph ×26,000. There are signs of intense pinocytotic activity at the membrane of the food vacuole. A number of secondary vacuoles have formed. In two of them one surface has collapsed inwards to meet the opposite surface and to give cup shaped profiles in section (arrowed). This form of collapse seems typical.

Fig. 5   Electron micrograph ×16,000. The membrane of a food vacuole at a moderately late stage. The irregularities of the membrane are signs of intense pinocytotic activity. The cup shaped profile of a secondary vacuole is arrowed.

Fig. 6   Electron micrograph ×15,000. A section through a food vacuole at a late stage when only undigested bacterial cell walls and some colloidal material are left. The vacuolar membrane has returned once more to a quiescent state. It is relatively smooth and without irregularities. This is typical of the appearance of a food vacuole immediately before it is expelled from the cytopyge (see Diagram 9).

130

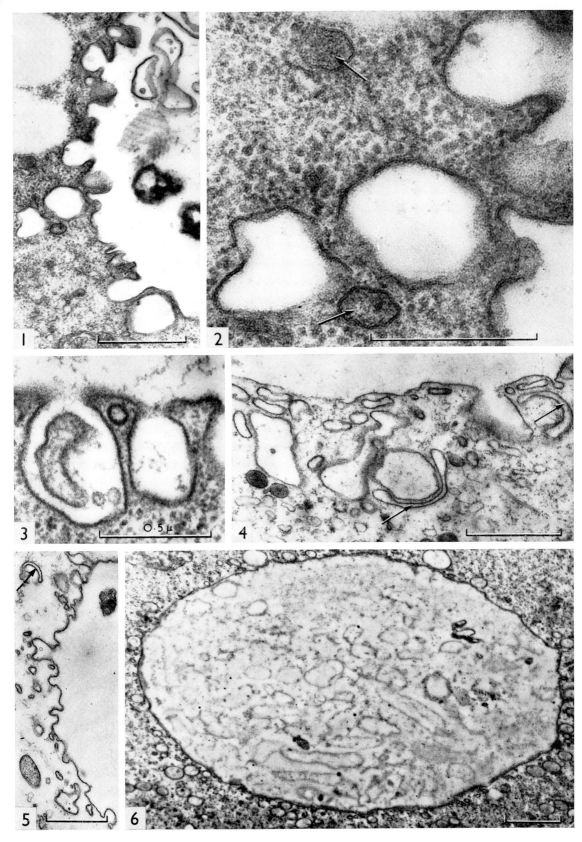

K

PLATE 25   Contractile vacuole   I

FIG. 1   Light micrograph ×2250. This section has been cut at right angles to the dorsal surface of the paramecium and shows a contractile vacuole in diastole when the vacuole is fully extended. The vacuole is discharged through a channel (dc) in the pellicle. On both sides of the contractile vacuole there are trichocysts which appear white.

FIG. 2   Electron micrograph ×10,000. This section was cut through the contractile vacuole at systole when the vacuole was nearly empty. The floor of the vacuole is almost in contact with its roof. The discharge channel (dc) in the pellicle is shaped like an inverted thimble with a closed septum which, in all the pictures shown here, separates the contractile vacuole from the water outside the paramecium.

FIG. 3   Electron micrograph ×20,000. An enlarged view of the discharge channel of a contractile vacuole in diastole. The channel occupies about as much space as the one corpuscular unit of the pellicle which it appears to replace, except that the channel penetrates deeply below the surface. Microtubules run just inside the surface of the channel and are here seen in cross-section. The same tubules extend down the wall of the vacuole and are seen in longitudinal section (see arrow).

FIG. 4   Electron micrograph ×8000. A similar section to Fig. 3 except that rather more of the contractile vacuole is shown. The contractile vacuole in diastole and systole is shown in Diagram 10, which was reconstructed from electron micrographs. At the lower right corner of the picture the contractile vacuole narrows to its connection with a nephridial canal (Plate 27).

FIG. 5   Electron micrograph ×8000. This section through the contractile vacuole at systole was cut parallel to the cell surface and therefore is at right angles to Fig. 2. The centre of the picture is occupied by the sectioned floor of the vacuole. The vacuole is represented only by a narrow irregular annulus which is surrounded by more cytoplasm outside the vacuole.

FIG. 6   Electron micrograph ×26,000. The section was cut at right angles to the cell surface through the discharge channel of a contractile vacuole, but not through the centre of the channel. Microtubules are clearly shown just inside the surface of the discharge channel and also running down from one side of the contractile vacuole.

132

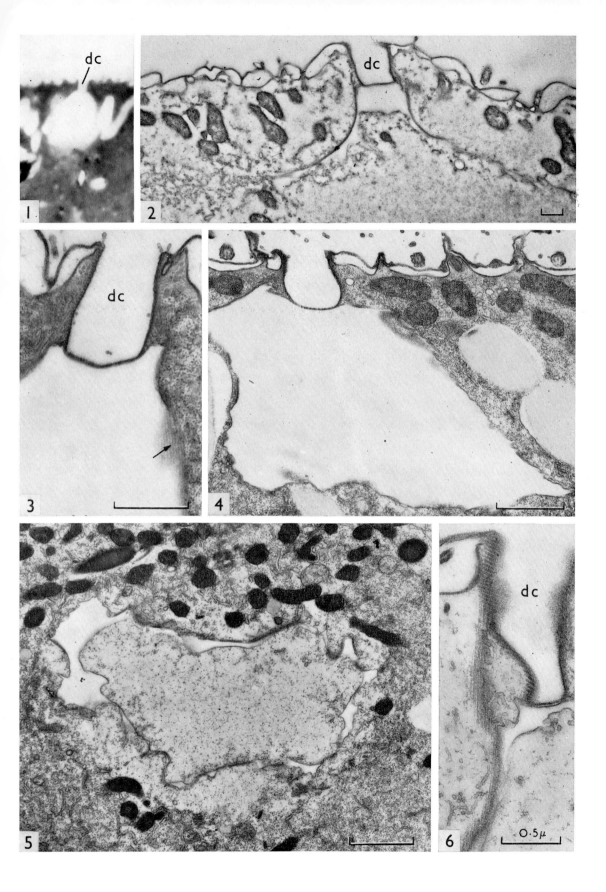

PLATE 26   Contractile vacuole II

FIG. 1   Electron micrograph × 36,000. The section was cut at right angles to the surface through the discharge channel (dc) of a contractile vacuole and its septum. There are some microvilli or small cytoplasmic protuberances near the top of the channel.

FIG. 2   Electron micrograph × 40,000. This section was cut parallel to Fig. 1 but almost tangentially to the surface of the discharge channel. The microtubules just inside the plasma membrane of the channel run in two directions. This can also be seen in Plate 25, fig. 6.

FIG. 3   Electron micrograph × 120,000. This shows the septum in transverse section as in Fig. 1 but at a higher magnification. At both the upper and the lower surface of the septum there is a unit membrane.

FIG. 4   Electron micrograph × 18,000. This section was cut parallel to the dorsal surface of the paramecium and through the lower part of the discharge channel. The section is viewed from outside the surface of the paramecium and shows seven ribbons of microtubules branching out from the channel. Each of these ribbons runs down the wall of the contractile vacuole (see Diagram 10 and Figs. 5 and 6) and then wraps itself round a nephridial canal (Plate 27).

FIG. 5   Electron micrograph × 17,000. This oblique section cuts the dorsal surface of the paramecium at the top of the picture and below there is a contractile vacuole at systole. Six of the microtubular ribbons are seen in section as they run down on the cytoplasmic side in contact with the membrane of the contractile vacuole. These ribbons run from the discharge channel to the nephridial canals as shown in Diagram 10.

FIG. 6   Electron micrograph × 32,000. This section was cut at a similar orientation and place to Fig. 5 and shows two of the microtubular ribbons (arrowed) in contact with the vacuolar membrane. Here the magnification is greater and the background cytoplasm is better preserved.

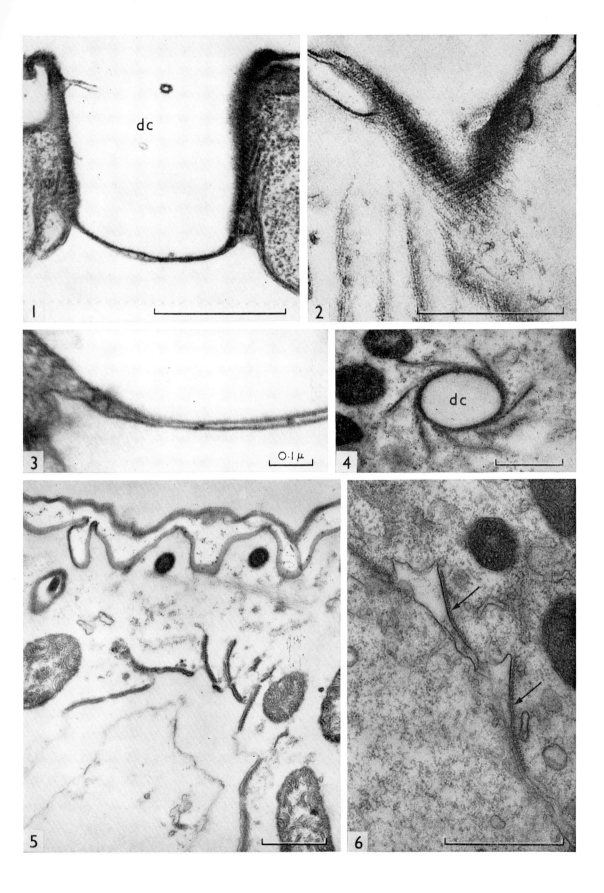

PLATE 27 The nephridial canals

The water, which is periodically expelled from the discharge channel of a contractile vacuole, has previously passed into the contractile vacuole from the nephridial canals. The nephridial canals collect the water from a network of minute nephridial tubules.

Fig. 1 Electron micrograph ×30,000. The nephridial canal has been transversely sectioned. The canal (c) is fully closed with opposite walls of the canal in contact with each other to give a flattened profile. The contractile vacuole is, however, in diastole at this stage. Nineteen microtubules can be seen in cross-section just on the cytoplasmic side of the membrane of the nephridial canal. Surrounding the canal there is a spongelike network of minute irregular nephridial tubules which do not appear to connect with the canal at this stage. This region of irregular tubules appears as a halo (just over 1 μ in diameter) round the canal and it is surrounded by normal cytoplasm of a denser and more granular appearance. This cytoplasm does, however, contain an orthotubular system (o) as in Plate 28, fig. 4.

Fig. 2 Electron micrograph ×18,000. This section shows part of a nephridial canal (c) and the structures in the surrounding cytoplasm. Much of the background cytoplasmic granules have been extracted, however, as can be seen in comparison with the other figures of this plate. The nephridial tubules (nt) round the canal are present in large numbers as are elements of the orthotubular system (o).

Fig. 3 Electron micrograph ×32,000. This longitudinal section through part of a nephridial canal (c) has been cut at right angles to Fig. 1, and at about the same stage. The microtubules round the canal membrane are shown in the plane of the section as fine parallel lines in a ribbon from the top to the bottom of the picture. The surrounding nephridial tubules (nt) are somewhat dimly seen against the cytoplasmic background.

Fig. 4 Electron micrograph ×32,000. An oblique section through an expanded nephridial canal (c) shows a surrounding mass of irregular nephridial tubules (nt) which at this stage appear to connect with the canal (arrows). Because the diameter of a tubule is less than the thickness of the section this cannot, however, be perfectly clear.

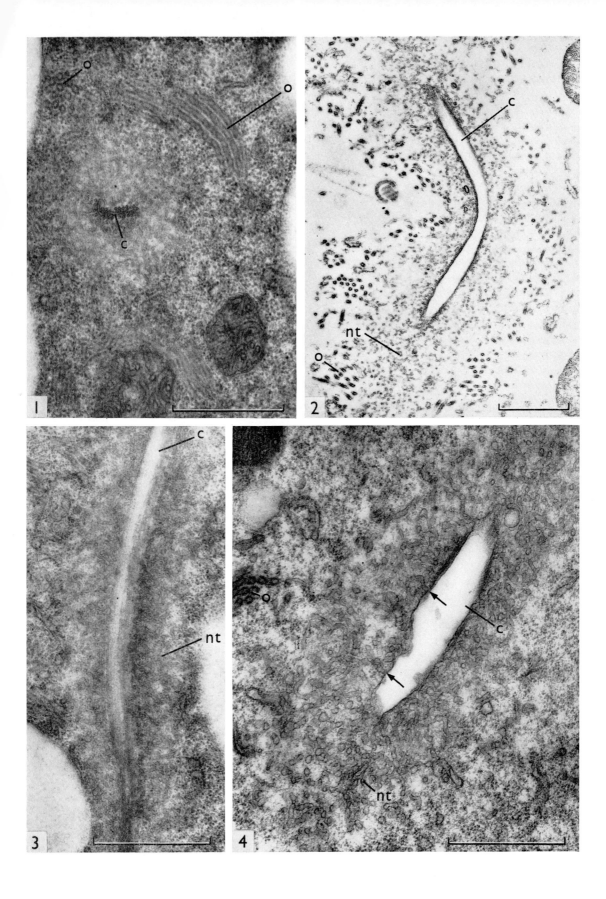

PLATE 28 Mitochondria and endoplasmic structures

In *Paramecium* the endoplasm is that part of the cytoplasm which is inside the level at which basal granules are found. The endoplasm contains organelles such as mitochondria, pretrichocysts and nuclei and various types of membrane complex such as endoplasmic reticulum as well as food storage granules like glycogen and lipid drops.

FIG. 1 Electron micrograph ×60,000. A mitochondrion is shown in longitudinal section and another is in transverse section. The mitochondria are bounded by two membranes, the innermost one being continuous with the tubular cristae which extend into the interior. The circular profiles to be seen inside the mitochondria represent transversely sectioned tubules. Within the matrix of one mitochondrion, there is a dense patch of material (arrowed) containing parallel fibrils. The cytoplasm surrounding the mitochondria contains granules a high proportion of which are probably glycogen granules.

FIG. 2 Electron micrograph ×32,000. The section shows a group of vesicles formed from smooth-walled membrane (arrowed). These may represent a vesicular type of Golgi complex as *Paramecium* does not have well-developed dictyosomes or Golgi complexes in the typical form of piles of flattened vesicles such as are found in vertebrates.

FIG. 3 Electron micrograph ×50,000. This complex of flattened smooth-walled membranes from the endoplasm of *P. aurelia* is the structure bearing the closest resemblance to a typical Golgi group. There is a similar structure near a trichocyst in Plate 42, fig. 5, but they are not abundant.

FIG. 4 Electron micrograph ×40,000. This section shows parallel groups of membranous tubules each 50 mμ in diameter and termed the orthotubular system. There are groups sectioned longitudinally, transversely and obliquely. The tubules show indications of a substructure in the form of a pattern of 10 mμ dots just outside the membranes. The orthotubular system is typically found in the cytoplasm near a contractile vacuole but its relationship to the vacuole is unknown. There are also a few vesicles of endoplasmic reticulum to be seen but if there are also free ribosomes present it is impossible to identify them with certainty.

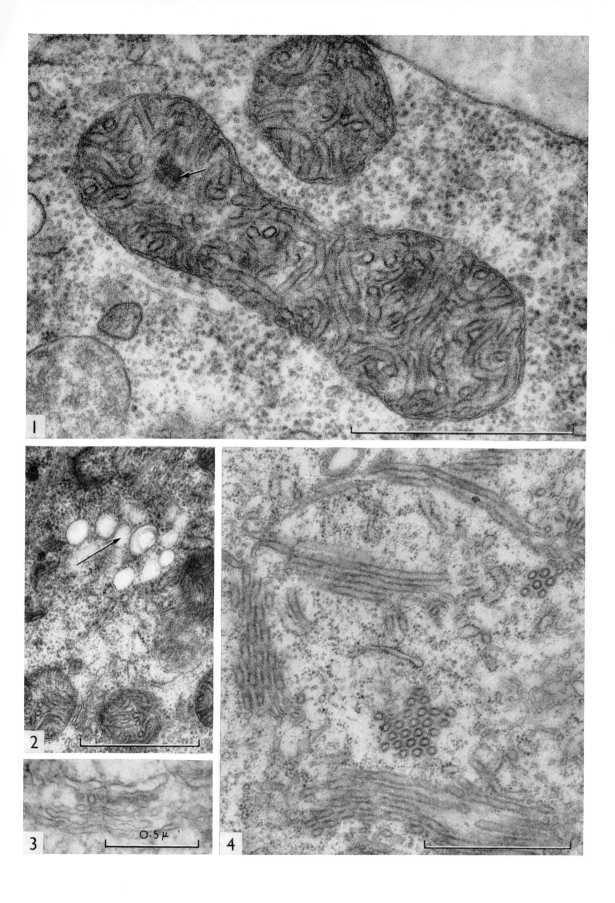

PLATE 29   Lipid drops, glycogen and ribosomes

FIG. 1   Electron micrograph × 40,000. Just above the mitochondrion there is an irregularly rounded body of a homogenous consistency. This is almost certainly a lipid droplet. Refractile lipid drops are found in cytoplasm freshly extracted from living paramecia (see Plate 51, figs. 1 and 2) but they are not so readily preserved in embedded material. Below the mitochondrion in the picture are some vesicles of rough endoplasmic reticulum (er) and these have rows of dark ribosomes on the outer surface of the membrane. The scale represents 0.25μ.

FIG. 2   Electron micrograph × 50,000. The section shows cytoplasm below a basal granule (b). Between the basal granule and the mitochondria there are a number of rounded dense dark grey glycogen granules (gl) 40 mμ in diameter. Near the bottom left corner of the picture some endoplasmic reticulum carries rows of ribosomes 15 mμ in diameter (arrowed). This micrograph was supplied by courtesy of Mr R. Sinden.

FIG. 3   Electron micrograph × 32,000. This shows glycogen taken from the cytoplasm of *Paramecium* which has been negatively stained with phosphotungstic acid. The glycogen granules appear white against the darker background.

FIG. 4   Electron micrograph × 120,000. This highly magnified section shows part of a mitochondrion with tubular cristae and two outer membranes. The cytoplasm shown beneath it has about thirty dense glycogen granules which are 40 mμ in diameter. There are also smaller grey granules in the background but it is not possible to estimate with confidence how many of these are free ribosomes. This micrograph is reproduced by courtesy of Mr Sinden and Miss M. M. Perry.

FIG. 5   Electron micrograph × 120,000. This section shows a region of cytoplasm similar to that shown in Fig. 4 and at the same magnification. This section, however, was treated with periodic acid and stained with lead citrate and shows a substructure consisting of particles 4 mμ in diameter within the granules which are 40 mμ in diameter. This reaction of granules to periodic acid treatment may be taken as characteristic of glycogen granules (Perry, 1967). Reproduction by courtesy of Sinden and Perry (unpublished).

140

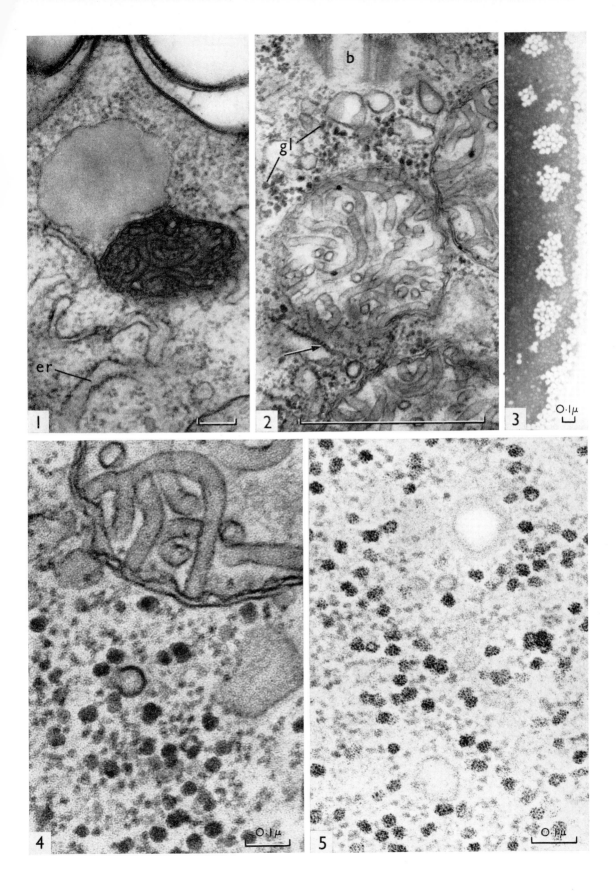

PLATE 30 Crystals

FIG. 1 Light micrograph ×780. This flattened whole-mounted paramecium, showing the ventral surface, has been silver stained and is in fact the same specimen as shown in Plate 2, fig. 1. This micrograph was taken by positive phase-contrast microscopy and the microscope was focused below the cell surface of the middle of the specimen where there were clusters of crystalline inclusions. The crystals are highly refractile and when observed by phase-contrast microscopy they become prominent because they cause phase reversal and glare. This is the reason for their very white appearance in the micrograph. Stocks of P. aurelia differ in the abundance of the crystals.

FIG. 2 Light micrograph ×280. This low power picture of a group of paramecia shows that the clusters of crystalline inclusions are readily seen at low magnification. The macronucleus in these animals is the only other obvious feature.

FIG. 3 Electron micrograph ×20,000. This section of cytoplasm near a food vacuole (fv) shows five crystalline inclusions (cr) which are each separated from the cytoplasm by a membrane. Each crystal therefore may be regarded as being in a separate vacuole. The crystals are harder than the cytoplasm and so are difficult to section. Sometimes they drop out of the section leaving a space and this seems to have happened in the case of the crystal to the lower left of the picture where the vacuole has a white interior. The crystals do not appear to have a regular shape.

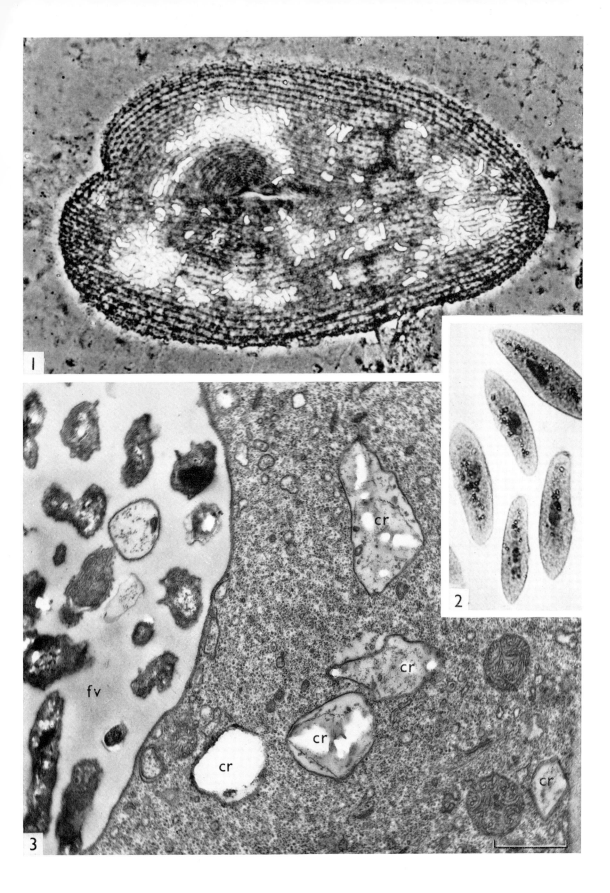

1

2

3

fv

cr

cr

cr

cr

cr

PLATE 31   Micronuclei and the macronucleus

FIG. 1   Electron micrograph ×22,000. A micronucleus in interphase is seen in section bounded by its nuclear envelope. There are three regions within the micronucleus. At the centre there is a very dense and compact osmiophilic granular region. Surrounding this there is a less dense and more disperse zone, apparently consisting of fibrils and granules. The remaining and major part of the micronucleus is of low density and appears to contain a network of fine fibres in a dispersed state. The dark central region cannot be a nucleolus because the micronucleus does not contain identifiable quantities of RNA. It may consist of condensed chromatin. Along the bottom of the picture is a small part of the macronucleus (ma).

FIG. 2   Electron micrograph ×14,000. The two micronuclei of a paramecium are both shown in section. They are both in interphase and they have a similar morphology to the micronucleus in Fig. 1 except that here the dense central region appears to have a less dense central core.

FIG. 3   Electron micrograph ×62,000. In this detail, part of the micronuclear envelope is seen in section to consist of two membranes which come together in places which probably correspond to nuclear pores.

FIG. 4   Light micrograph ×2250. The micronuclei are usually found close to the macronucleus. Here the micronucleus (mi) has the macronucleus (ma) on three sides of it. At the top left corner is part of the gullet (g).

FIG. 5   Electron micrograph ×24,000. The section shows a micronucleus (mi) close to the macronucleus (ma) as in Fig. 4. The nuclear envelope which bounds the macronucleus has an irregular shape; there are large and small dense bodies in the macronucleus.

144

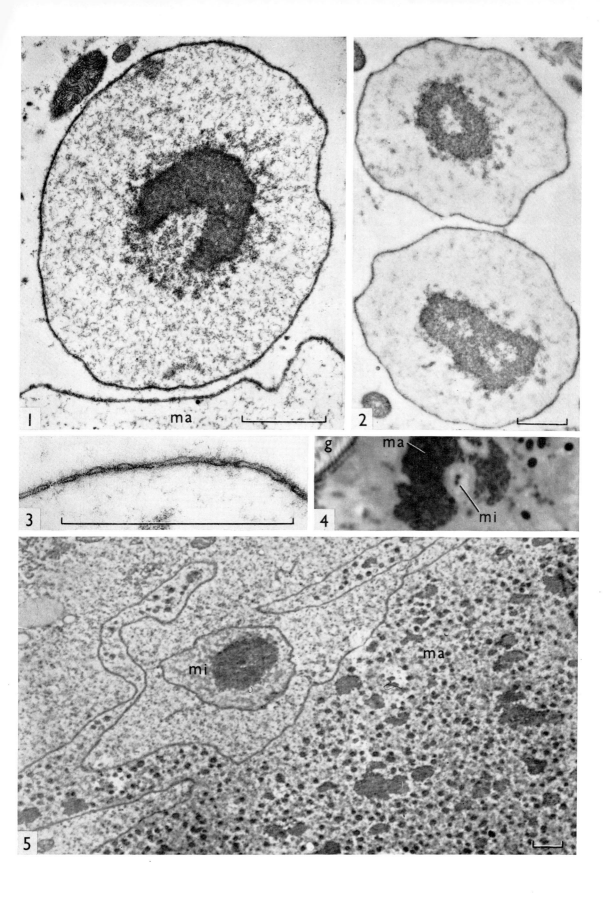

PLATE 32   The macronucleus

FIG. 1   Electron micrograph × 5600. Most of this picture is occupied by a macronucleus seen in section. The great size of the macronucleus compared with a micronucleus can also be seen in Plate 14, fig. 1. In the macronucleus of *Paramecium* there are two classes of dense bodies which differ in size, the larger being 0.5–1 μ in diameter and the smaller being 0.1–0.2 μ in diameter. The small bodies are more numerous than the large bodies.

FIG. 2   Electron micrograph × 4500. This section shows five macronuclei isolated *in vitro* by the method of Stevenson (1967). The nuclear membrane seems to have been lost but despite this the macronuclei seem to have held themselves together. Large and small bodies can be seen.

FIG. 3   Electron micrograph × 26,000. This is an oblique section through part of the nuclear membrane of a macronucleus of *P. aurelia*. A number of the pores in the nuclear membrane are seen in the plane of the section as dark rings about 80 mμ in diameter round a less dense central area which is 40 mμ in diameter. In some cases there is a small dense granule about 20 mμ in diameter in the middle of the pore. Small bodies can be seen in the macronucleus in the top half of the picture.

FIG. 4   Electron micrograph × 63,000. This detail shows part of the nuclear envelope of the macronucleus in transverse section. The envelope consists of two membranes separated by a less dense gap but the two membranes come together at regular intervals in places which correspond to the pores of the nuclear membrane. The macronucleus is at the top of the picture.

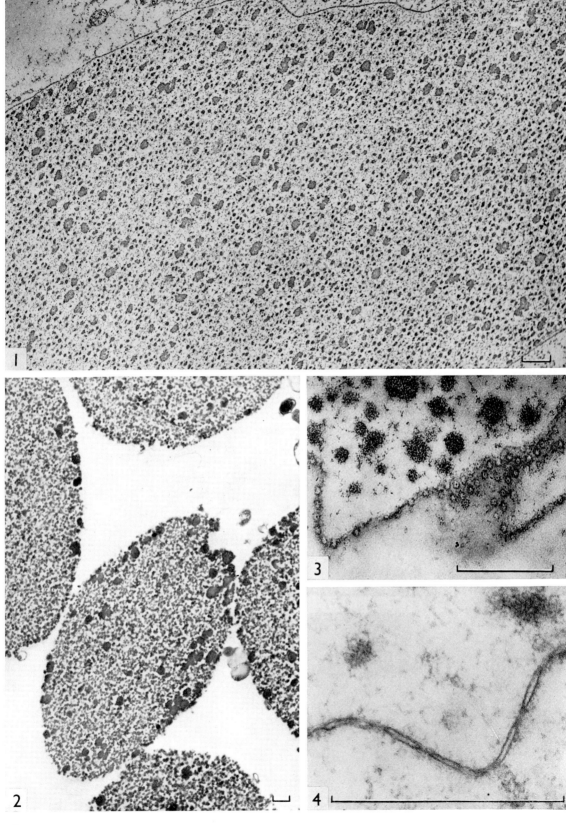

L

PLATE 33   Large bodies in the macronucleus

Fig. 1   Electron micrograph × 50,000. One large dense body about 1 μ in diameter and over thirty small dense bodies which are 0.1–0.2 μ in diameter are shown in this detail of part of a macronucleus. The large body characteristically shows a central region of lesser density. Large bodies probably correspond to nucleoli as they have been shown to contain RNA. Between the large and small bodies there is a network of irregular fine fibrils. The nuclear envelope runs across the lower half of the picture and is seen in section. The inner and outer membranes of the nuclear envelope are clearly seen. There are glycogen granules and smaller particles in the cytoplasm at the bottom of the picture. This photograph is reproduced by courtesy of Miss M. M. Perry.

Fig. 2   Electron micrograph × 80,000. This detail of part of a macronucleus shows a large dense body and some small dense bodies and between which is irregular fibrillar material. The large body appears to have particulate or granular ultrastructure in contrast to the small bodies which have a more fibrillar ultrastructure of the same nature and probably continuous with the fibrils between the small bodies. The small bodies may represent chromatin in the condensed state and the fibrillar material between them may represent chromatin in the expanded state. To the upper right of the large body and just in contact with it there is a dense body (see arrow) of similar size to the small bodies but whose granular ultrastructure shows it to be part of the large body.

Fig. 3   Electron micrograph × 33,000. The three large dense bodies shown in this detail of a macronucleus each have a prominent central region of lesser density. To the bottom of the picture there is a group of rather irregular microtubules (arrowed) of a type rarely found in macronuclei. At amitosis macronuclei contain ordered microtubules (see Plate 37).

148

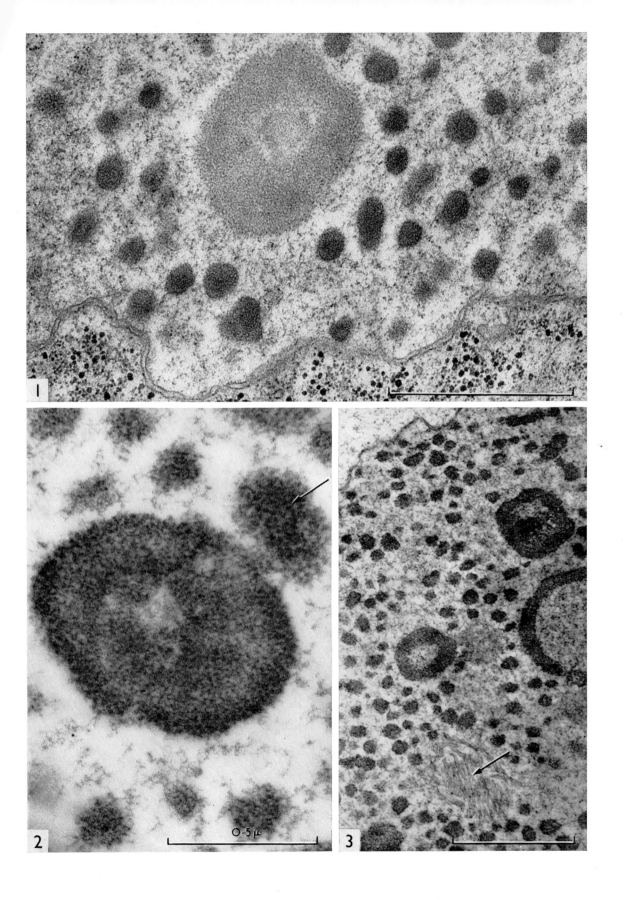

0·5μ

PLATE 34   Micronuclei at mitotic metaphase

FIG. 1   Electron micrograph ×20,000. The two micronuclei of *P. aurelia* divide by mitosis just before fission takes place. Here one of the micronuclei is seen in section at metaphase in an equatorial view. The plane of the metaphase plate runs from just above the bottom left corner to the top right corner of the picture and the spindle poles are to the top left and the bottom right of the nucleus as it is seen here. The nuclear membrane is obviously intact and it remains so during mitosis which of course is quite different to the situation in a metazoan. Otherwise the appearance of this metaphase is conventional. The chromosomes appear as dense masses with localised centromeres, or kinetochores, where groups of microtubules are joined to the chromosomes. Other microtubules run between the chromosomes from pole to pole of the spindle. The microtubules all run in approximately polar directions although the poles of the spindle are not strongly localised at a point but occupy an area of nuclear membrane in the polar region. Part of the macronucleus is to be seen to the right of the picture.

FIG. 2   Electron micrograph ×23,000. This is the other micronucleus from the same section of the same paramecium as in Fig. 1. Although the two micronuclei pass through mitosis at the same time they are not completely synchronous in this case. The nuclear membrane is slightly elongated in the polar direction in which the microtubules run and the chromosomal material is more scattered about the equatorial plane than was the case in Fig. 1. Probably this is an early anaphase stage.

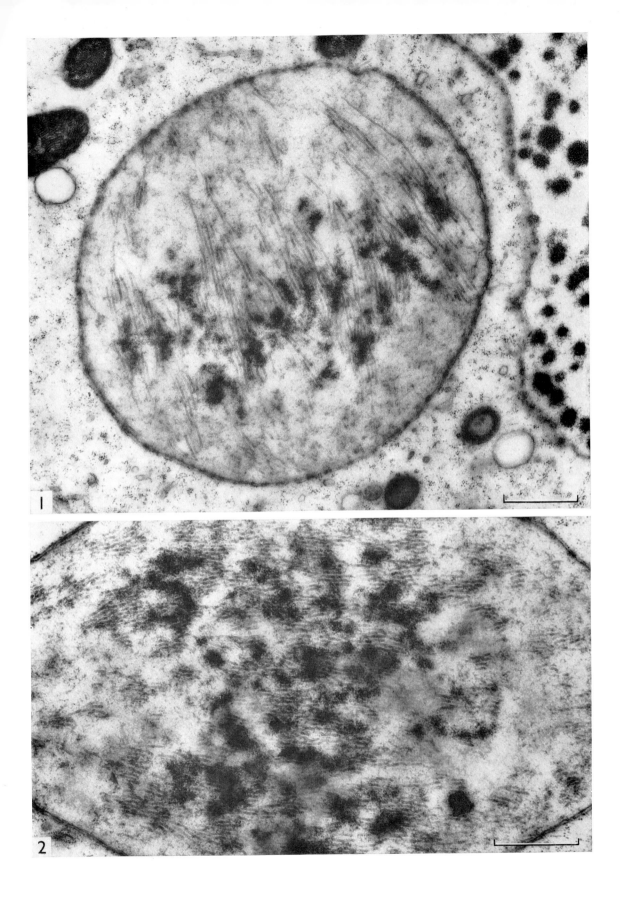

1

2

PLATE 35   Micronucleus in mitosis

FIG. 1   Electron micrograph ×45,000. A part of a micronucleus is shown. There is a central ring of high density condensed chromatin surrounded by a region of low density. This is characteristic of the distribution of chromatin in interphase and is similar to that shown in Plate 31, figs. 1 and 2. There is a layer of microtubular material (arrowed) immediately inside the nuclear membrane. It seems probable that the absence of a similar layer of microtubules from Plate 31, figs. 1 and 2 is due to the use of the methacrylate embedding method in these cases. The microtubular layer is most probably present throughout interphase.

FIG. 2   Electron micrograph ×24,000. This micronucleus has its condensed chromatin more widely distributed throughout the nuclear volume than is the case in Fig. 1. This micronucleus is therefore judged to be in prophase. There is a layer of microtubular material (arrowed) within the nuclear membrane and there are regions of high, intermediate and low density chromatin.

FIG. 3   Electron micrograph ×51,000. This detail is an enlargement of a small area of Plate 34, fig. 1. It shows microtubules attached to a localized region of a metaphase chromosome which is represented by the high density material.

FIG. 4   Electron micrograph ×60,000. This is a small part of a telophase micronuclear spindle sectioned longitudinally in a region between the daughter groups of chromatin. The nuclear envelope is intact and is represented by a pair of membranes separated by a gap of about 5 mμ. Within the nuclear envelope there is a stem body consisting of closely packed parallel microtubules.

FIG. 5   Electron micrograph ×30,000. This is also part of a telophase spindle in longitudinal section and although a greater length of it is shown than in Fig. 4, it is much longer in its entirety. A fragment of the orthotubular system (o) associated with contractile vacuoles is shown in the cytoplasm outside the nuclear membrane.

152

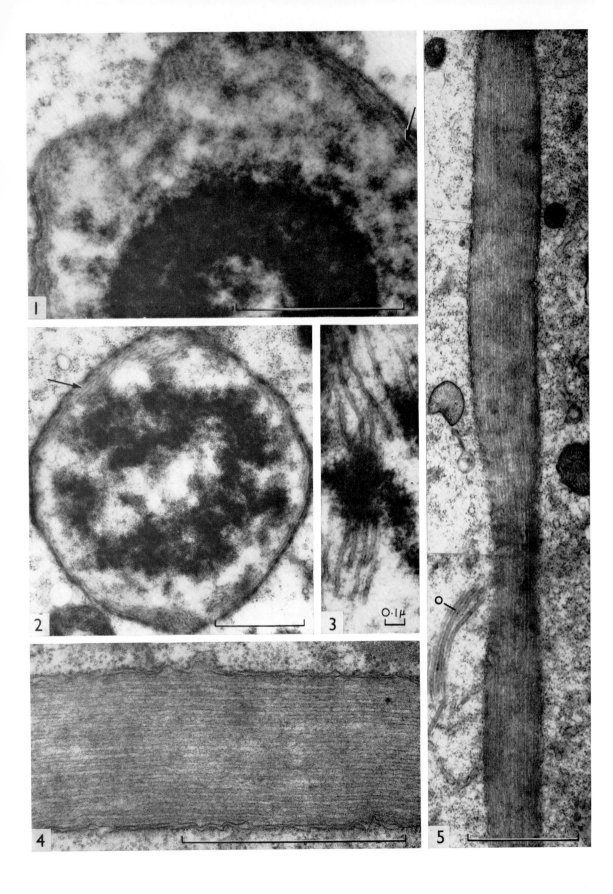

O.1μ

PLATE 36   The fission furrow

FIG. 1   Electron micrograph × 20,000. This longitudinal section of a paramecium in fission demonstrates the region of the fission furrow. There is a narrow neck or isthmus of cytoplasm which still joins the daughter paramecia. There is no sign of dedifferentiation of the pellicle structures at the surface of the furrow region (at the bottom left and top right of the picture). Endoplasmic reticulum (er) carrying ribosomes is found between the mitochondria in the cytoplasm near the furrow. Immediately below the furrow there are dense accumulations of fine fibrillar material (f), but it is also normal to find these fibrils at this level below the pellicle in paramecia not in fission. The daughter paramecium to the bottom right of the picture is the proter, because a kineto-desmal fibril (k) can be seen to run in this direction from its basal granule. One juvenile trichocyst (jt) can be seen and also a transversely sectioned ribbon of six microtubules which runs just below the surface of a longitudinal ridge separating two kineties. Microtubules have been observed in this position only in paramecia during fission.

FIG. 2   Electron micrograph × 50,000. This detail is an enlargement of part of Fig. 1. It shows more closely the ribbon of microtubules (arrowed) below a longitudinal ridge. The membranes of the pellicle are also clearly shown. The plasma membrane has microvilli (mv).

154

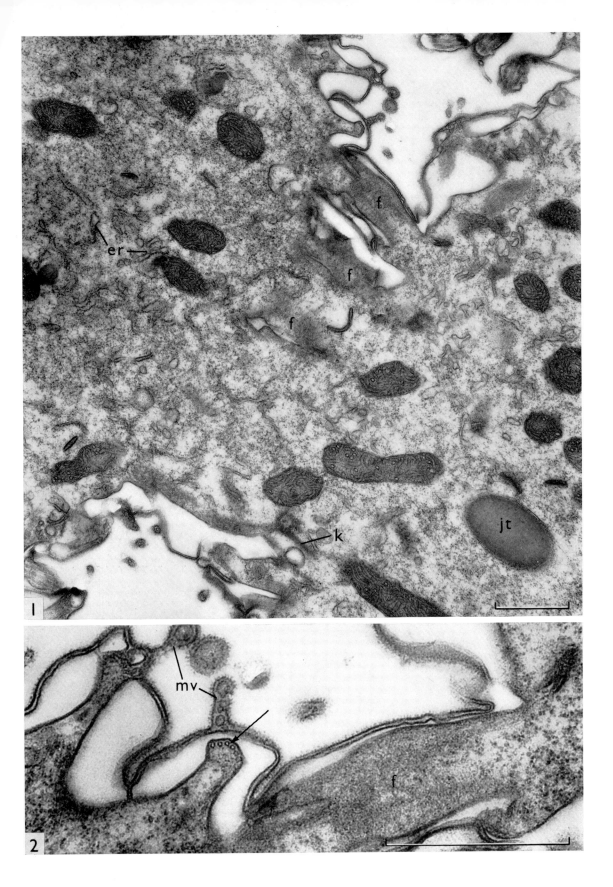

PLATE 37   Amitosis of the macronucleus

FIG. 1   Light micrograph ×250. This Feulgen-stained, whole-mounted specimen of *P. aurelia*, which is at an earlier stage of fission than that in Plate 36, shows the elongated and constricted shape of the macronucleus during amitosis.

FIG. 2   Light micrograph ×1800. The longitudinal section through a paramecium in fission has been stained with toluidine blue. It shows a similar stage to Fig. 1. The macronucleus which contains darkly stained large bodies is constricted in the furrow region and has a shape like an hour glass. There are two food vacuoles to the left of the picture.

FIG. 3   Electron micrograph ×12,000. This section has been cut at a similar orientation to Fig. 2 through a paramecium at a similar stage of the fission process. Part of the fission furrow can be seen in the top left corner of the picture. Only the constricted isthmus is shown between the two main masses of the macronucleus. At a slightly later stage the macronucleus pinches into two macronuclei in this region. The nuclear membrane remains intact. Large and small bodies can be seen in the macronucleus and there are also microtubules which run between the dense bodies in the direction in which macronucleus is elongated.

FIG. 4   Electron micrograph ×40,000. This is from a thin section near that of Fig. 3, but at a higher magnification. It shows part of the isthmus of the macronucleus and groups of microtubules (arrowed) can be seen to run between the dense bodies. There is a fine fibrous background between the dense bodies.

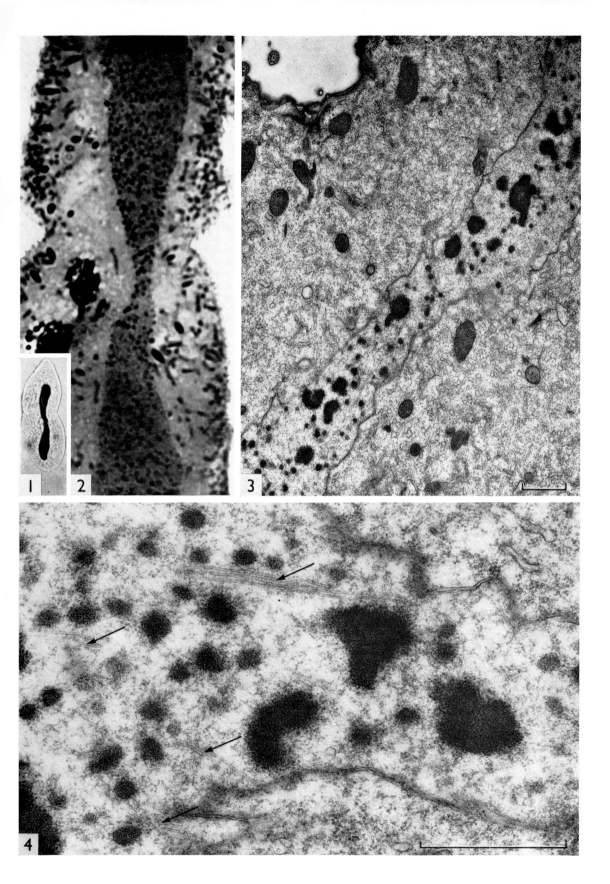

PLATE 38 The formation of new surface structures during fission

This plate shows features of the pellicle from paramecia in which the fission furrow was present at an early and later stage of its development. The figures were taken neither from the furrow region nor from the polar regions but from intermediate regions where extensive growth of the pellicle took place.

FIG. 1 Electron micrograph ×40,000. A group of four basal granules is shown in a section cut transversely to the surface of the pellicle. Two basal granules bear cilia and formerly they formed a pair at the centre of one corpuscular unit of the pellicle. Two new basal granules have formed in a position anterior to the old basal granules, that is as shown here to the right of the basal granules with cilia.

FIG. 2 Electron micrograph ×40,000. A group of four basal granules similar to those shown in Fig. 1, but in a section cut transversely to the basal granules.

FIG. 3 Electron micrograph ×26,000. A longitudinal section shows the basal granules of a kinety with the anterior direction to the left of the picture. The basal granules are in pairs and each pair consists of a newly formed as well as an old basal granule. The old basal granules show attachment to the cell surface but the new basal granules do not and in some cases they are not oriented in parallel with the old basal granule. The pairs of basal granules have probably been derived from groups of four as in Figs. 1 and 2.

FIG. 4 Electron micrograph ×12,000. This section has been cut rather more obliquely to the cell surface than was the section in Fig. 3 and it shows a kinety at a slightly later stage. The anterior direction is to the left. The basal granules are still in pairs but growth of the cell and the membranes of the pellicle has increased the separation between adjacent pairs.

FIG. 5 Electron micrograph ×32,000. An oblique section through part of a ridge between adjacent kineties shows vesicular cytoplasm (arrowed) immediately below the plasma membrane. The striations below the vesicles indicate microtubules.

FIG. 6 Electron micrograph ×80,000. A longitudinal section through part of a similar ridge to that shown in Fig. 5 shows vesicles below the plasma membrane and microtubules (arrowed) below the vesicles.

FIG. 7 Electron micrograph ×80,000. This detail from a similar section to Figs. 5 and 6 shows a microvillus in longitudinal section at the cell surface. The wall of the microvillus has a plasma membrane with a thin diffuse layer exterior to it. Microvilli and vesicles are characteristic of the cell surface when the pellicle is in a state of rapid growth.

158

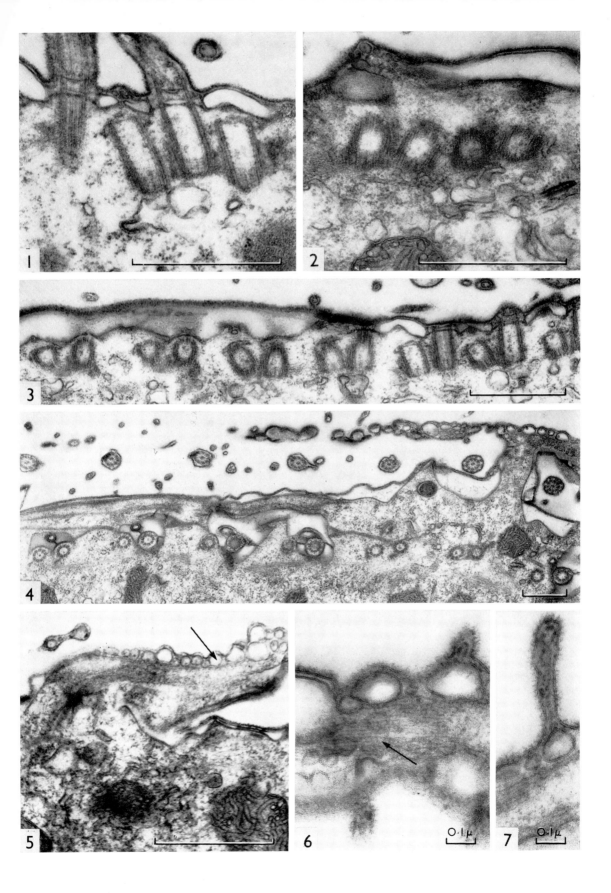

PLATE 39   A new gullet at the fission stage

FIG. 1   Light micrograph × 620. This longitudinal section of a paramecium in fission has been stained with toluidine blue. The fission furrow can be seen. In both the future anterior animal or proter (to the right) and the posterior animal or opisthe, there is a gullet (g). The newly formed gullet (arrowed) of the opisthe appears as an invagination of the cell surface while the original gullet has passed to the proter.

FIG. 2   Electron micrograph × 32,000. This is from a similar section to Fig. 1 and shows part of a newly formed gullet but the position of the structures within the gullet was not determined. Dense bundles of parallel microtubules run for a considerable distance into the cytoplasm from the lower surface of a row of basal granules. Comparable structures have not been observed in *Paramecium* between fissions and it is not possible to say whether these microtubules are a transient feature of gullet development or whether this is the anterior extremity of the postesophageal fibres shown in Plate 19.

FIG. 3   Electron micrograph × 30,000. This is part of a section adjacent to Fig. 2. The basal granules indicated with arrows are identical to those arrowed in Fig. 2. The comparison of serial sections enables a structure to be reconstructed in three dimensions. In this case certain basal granules clearly sectioned in this figure appear only at grazing incidence in the section of Fig. 2.

FIG. 4   Electron micrograph × 40,000. This detail is from the same series of sections shown in Figs. 2 and 3. It shows a basal granule in longitudinal section with microtubules attached to its lower surface.

FIG. 5   Electron micrograph × 19,000. This detail from the endoplasm of a cell in fission shows the profiles of two mitochondria each apparently with a constriction like an equatorial furrow. Such figures support the idea that mitochondria increase in number by the growth and subsequent fission of a previous generation of mitochondria.

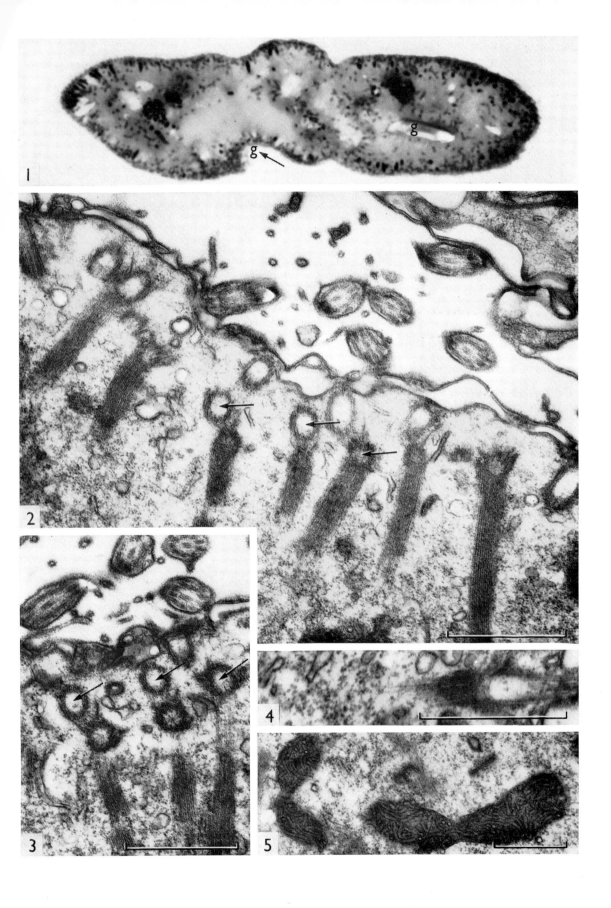

PLATE 40 Mitochondrial replication by budding and development

Mitochondria may duplicate by fission (see Plate 39, fig. 5) but Wohlfarth-Bottermann (1957) has suggested that in *Paramecium* they originate from undifferentiated vesicles. This plate presents observations which seem to show that the vesicles are formed in contact with mature mitochondria and then develop into new mitochondria. It is suggested that Figs. 1 to 6 trace the sequence of development.

Fig. 1 Electron micrograph ×60,000. The membrane-bounded vesicle (arrowed) has a moderately dense finely granular matrix. Its membrane seems to be continuous with the outer membrane of the mitochondrion to the right of the figure with which it is in contact.

Fig. 2 Electron micrograph ×40,000. A similar but larger membrane bounded vesicle is in contact with a mitochondrion.

Fig. 3 Electron micrograph ×60,000. Three undifferentiated membrane bounded vesicles similar to those of Figs. 1 and 2 are shown in the cytoplasm and are apparently separate from the mitochondrion shown to the left of the figure. This kind of vesicle can be found only amongst mature mitochondria in the cortical region of the *Paramecium* cell.

Fig. 4 Electron micrograph ×60,000. A separate vesicle (arrowed) bounded by a unit membrane, has within its finely granular matrix a number of denser tubular structures which are considered to be the precursors of the tubular cristae of a mitochondrion. A mature mitochondrion is shown above the vesicle.

Fig. 5 Electron micrograph ×40,000. This shows a vesicle considered to be at a later stage of its differentiation into a mitochondrion. The tubular cristae are well formed and appear to be in contact with a membrane which encloses the tubular region and which has formed within the outer membrane of the vesicle. Inside the differentiated vesicle there is low density background material which is neither membrane nor tubule.

Fig. 6 Electron micrograph ×40,000. This figure shows a mitochondrion (arrowed) which is considered to be in the last stage of the developmental process. It differs from mature mitochondria, a number of which appear in the figure, only in that its matrix material is of a somewhat lower density. Between the central mitochondrion and the one shown below it, rough endoplasmic reticulum (er) can be seen.

162

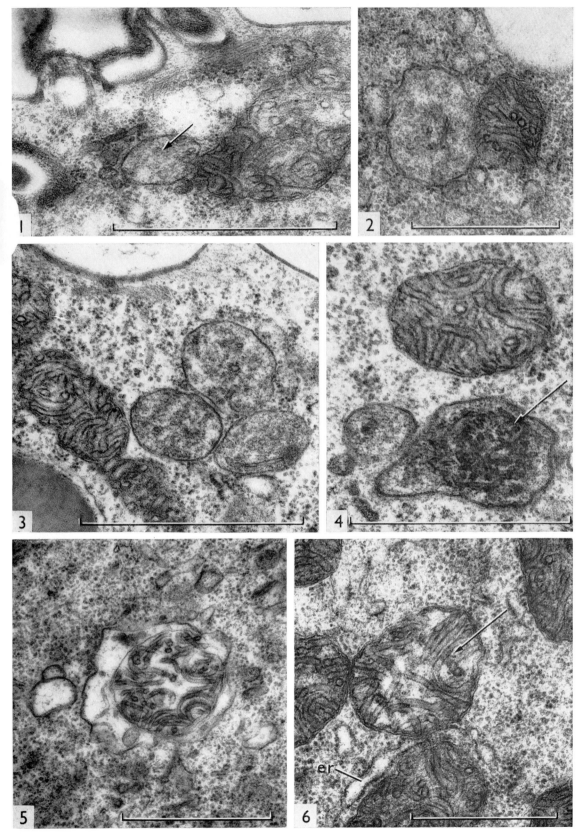

M

PLATE 41   Trichocysts and the fission cycle

Fig. 1   Electron micrograph × 12,000. A paramecium has been transversely sectioned near its anterior or posterior end. The white rounded objects in the central area of the section are the bodies of mature trichocysts seen in transverse section. It is obvious that trichocysts can occupy a considerable volume within a paramecium. This paramecium is probably about midway between successive fissions because no immature trichocysts can be seen.

Fig. 2   Electron micrograph × 7500. A section through another paramecium shows the bodies of both mature and immature trichocysts below the pellicle. Mature trichocysts appear white but the immature or juvenile trichocysts (arrowed) appear dark and dense because they are osmiophilic. A considerable proportion of juvenile trichocysts co-existing with many mature trichocysts is characteristic of a paramecium in the second half of the interfission period.

Fig. 3   Light micrograph × 1500. A longitudinal section through a paramecium has been cut parallel with the nearest cell surface to show the bodies of trichocysts in transverse section. The section was stained with toluidine blue. Mature trichocysts are unstained and appear white whereas the juvenile trichocysts appear black.

Fig. 4   Light micrograph × 1500. Another longitudinal section cut at a deeper level than Fig. 3 shows both mature and juvenile trichocysts as elongated bodies below the pellicle. The mature trichocysts appear unstained by the toluidine blue method and juvenile trichocysts are darkly stained and appear black.

Fig. 5   Electron micrograph × 12,000. Part of a similar section to Fig. 4 shows the bodies of mature trichocysts which appear white together with juvenile trichocysts (arrowed) which are osmiophilic and smaller than the mature trichocysts.

Fig. 6   Light micrograph × 1500. This longitudinal section stained with toluidine blue shows one half of a paramecium in fission, the fission furrow being to the bottom left of the picture. The trichocysts below the pellicle are all juvenile and appear black except for a group (arrowed) at the pole of the paramecium which appear white and are therefore mature trichocysts.

Fig. 7   Light micrograph × 2200. This detail is an enlarged view of one pole of the paramecium shown during fission in Plate 39, fig. 1. Both mature and immature trichocysts can be seen.

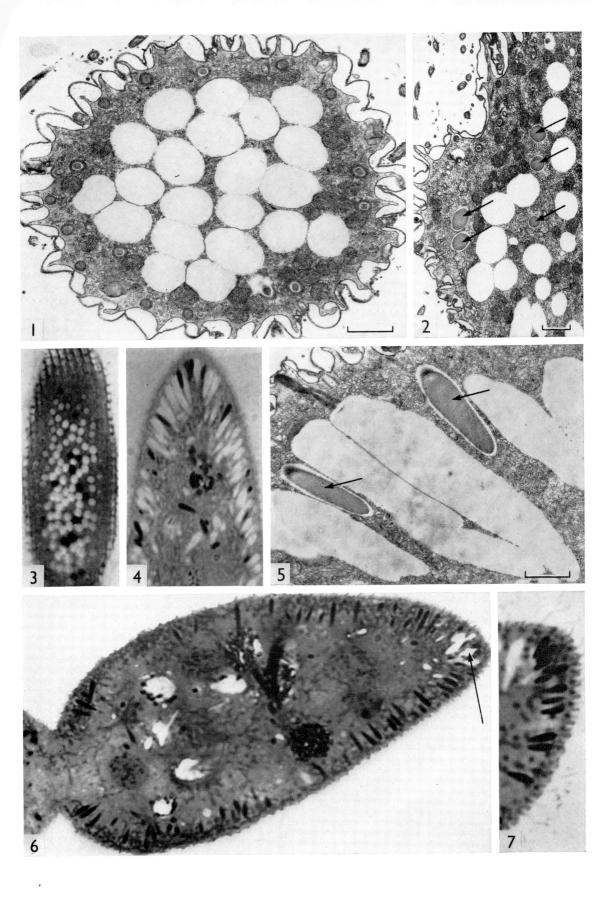

PLATE 42 Pretrichocysts and the development of trichocysts

FIG. 1 Electron micrograph × 32,000. An early pretrichocyst stage in the endoplasm. Trichocysts develop from small undifferentiated vesicles in the endoplasm. In the present stage the vesicle has enlarged and there is a dense oval crystalline central zone surrounded by a low density region which appears to contain fine irregular fibrils. A single membrane encloses the pretrichocyst.

FIG. 2 Electron micrograph × 80,000. An enlargement of part of Fig. 1 shows more clearly the periodicities in the crystalline central zone of the pretrichocyst. The dark lines appear at intervals of 11 mμ and probably represent layers of closely packed macromolecular units.

FIG. 3 Electron micrograph × 24,000. The section shows a pretrichocyst at a later stage of development. The vesicle has enlarged and the dense crystalline central zone has assumed an elongated shape which is already similar to the body of the mature trichocyst.

FIG. 4 Electron micrograph × 30,000. A late pretrichocyst stage is shown in the endoplasm of a paramecium. It is still bounded by a membrane, now elongated to accommodate the lengthened central crystalline zone which has developed a trichocyst tip in continuity with the trichocyst body.

FIG. 5 Electron micrograph × 15,000. The section shows a row of four juvenile or immature trichocysts at the particular places in the pattern of pellicle ultrastructure which they may occupy. Since the pretrichocyst stage shown in Fig. 4, they have migrated from the endoplasm to their sites in contact with the pellicle and are therefore no longer termed pretrichocysts. Development of the trichocyst sheath to enclose the tip of the trichocyst has also taken place and the low density region between the dense crystalline region and the limiting membrane has been reduced until it is no longer present.

FIG. 6 Electron micrograph × 60,000. The section shows what appears to be the remnant of a trichocyst (arrowed) which has degenerated in the endoplasm. A pretrichocyst (pt) and the tip of another trichocyst (tt) can also be seen.

166

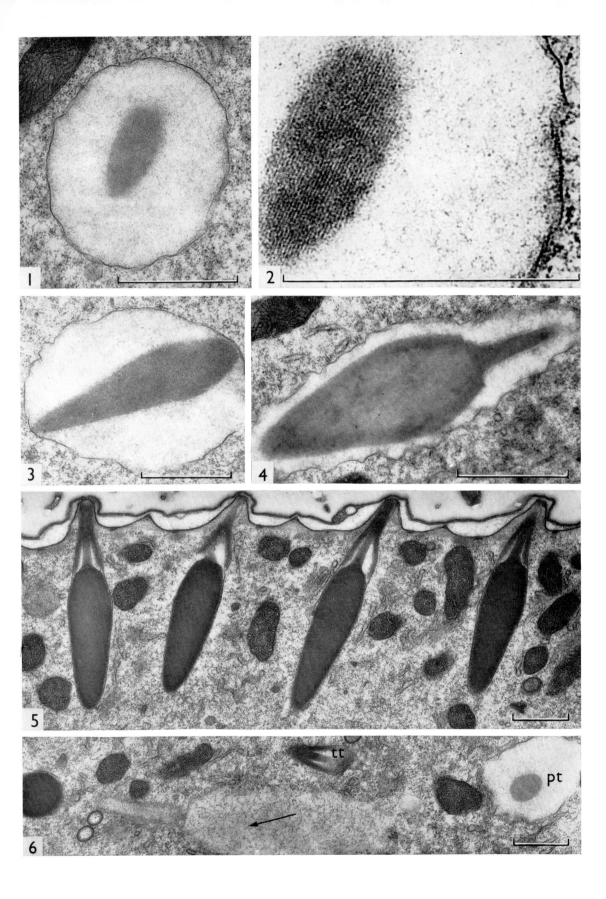

PLATE 43   Ultrastructure of the juvenile trichocyst

Fig. 1   Composite electron micrograph ×62,000. This figure of a longitudinal section through a juvenile trichocyst is made up of three joined prints. The whole length of the body of the trichocyst is shown and the base of the trichocyst tip is to the top of the picture. The whole trichocyst body appears to be formed of one dense crystalline lattice structure. Outside the trichocyst body there is a single layer of dense spherical units (13 mμ in diameter) and the limiting membrane runs immediately outside this layer.

Fig. 2   Electron micrograph ×97,000. The section shows another juvenile trichocyst in the region of the join between the body and the tip of the trichocyst. At this join there is a shoulder-like change in the outline of the trichocyst but no obvious discontinuity or change in the crystalline ultrastructure. The crystalline lattice appears as parallel dark lines at intervals of 8 mμ; two other sets of lines inclined at about 65° to the first set can be seen towards the edges of the trichocyst. This pattern is somewhat similar to the crystalline ultrastructure of amphibian yolk platelets as seen by electron microscopy which suggests that there are macronuclear units within juvenile trichocysts in some hexagonal regularly packed arrangement.

Fig. 3   Electron micrograph ×150,000. This is an enlargement from the lower part of Fig. 1, to show the ultrastructural detail more clearly. The dense parallel lines are at intervals of 6 mμ. It is remarkable that in the change from a juvenile to a mature trichocyst (as shown in Plate 12) the juvenile trichocyst loses its affinity for osmium tetroxide or stain and will appear structureless (see also Plate 41).

168

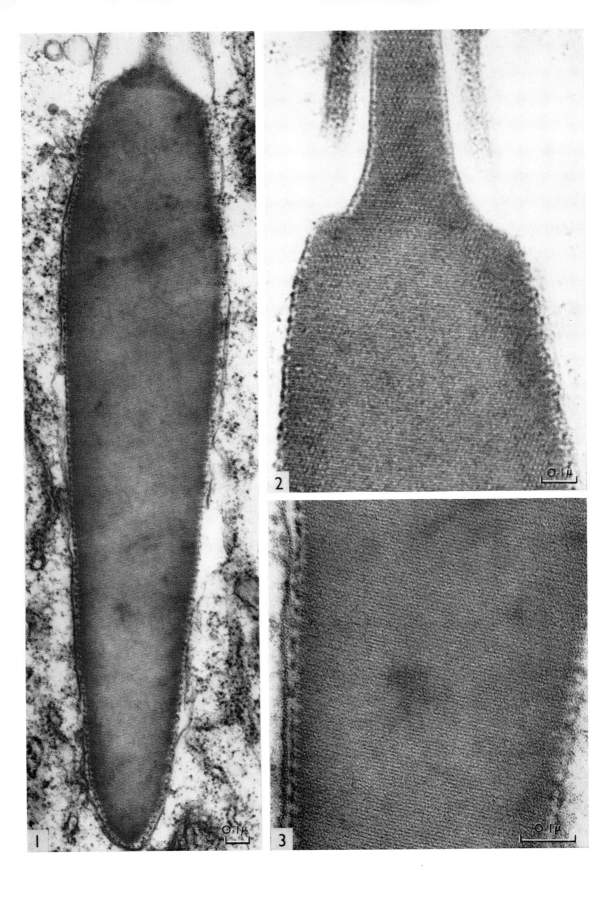

PLATE 44   Meiosis during conjugation

These light micrographs of stained whole-mounted specimens of *P. aurelia* during conjugation are reproduced by courtesy of Dr K. W. Jones (Jones, 1956). In Figs. 2 to 7 the preparations were stained with Azure-A after hydrolysis so that the stain would indicate the distribution of DNA as in the Feulgen method (DeLamater, 1951).

FIG. 1   Light micrograph ×2500. The two micronuclei are in interphase and were fixed and stained with toluidine blue before the occurrence of any nuclear change induced by conjugation. There is a dense ring-shaped chromatin area in the middle of the faintly stained nuclei. This figure should be compared with the two similar micronuclei shown by electron microscopy in Plate 31, fig. 2.

FIG. 2   Light micrograph ×3300. The two micronuclei are shown at a very early stage of the first meiotic division, about 30 min after the animals of complementary mating types were mixed. The chromatin stained with Azure-A is more uniformly distributed in the nucleus here than in Fig. 1, but two small dark granules can just be discerned within it.

FIG. 3   Light micrograph ×4000. One of the two micronuclei shows the extrusion of a strand of chromatin (arrowed) from the main mass of chromatin. At the point where this strand joins the main nuclear mass a dark granule of condensed chromatin (an x-body of Jones, 1956) can be seen. The extrusion of chromatin threads marks the beginning of the crescent stage of meiotic prophase.

FIG. 4   Light micrograph ×1950. Two micronuclei are shown with their chromatin extended in a typical crescent stage. The macronucleus is shown just below the micronuclei. The crescent stage occurs between 1 and 2 h after the beginning of conjugation (see also Diagrams 12 and 13).

FIG. 5   Light micrograph ×4700. This shows a crescent stage of meiotic prophase in which fine chromatin strands extend between two regions of condensed chromatin at each end of the nuclei.

FIG. 6   Light micrograph ×4750. The two micronuclei are shown in metaphase of the first meiotic division. The metaphase plates are viewed equatorially. In each nucleus there is a dense mass of condensed chromosomes within a spindle-shaped nuclear envelope.

FIG. 7   Light micrograph ×4500. The chromosomes are from a squash preparation of a micronucleus in metaphase of the first meiotic division. In the top left corner of the figure some metaphase chromosomes from the other micronucleus can be seen. This preparation was from *P. aurelia*, stock 144, syngen 1, and the diploid chromosome number was estimated at eighty-eight or ninety by Jones (1956).

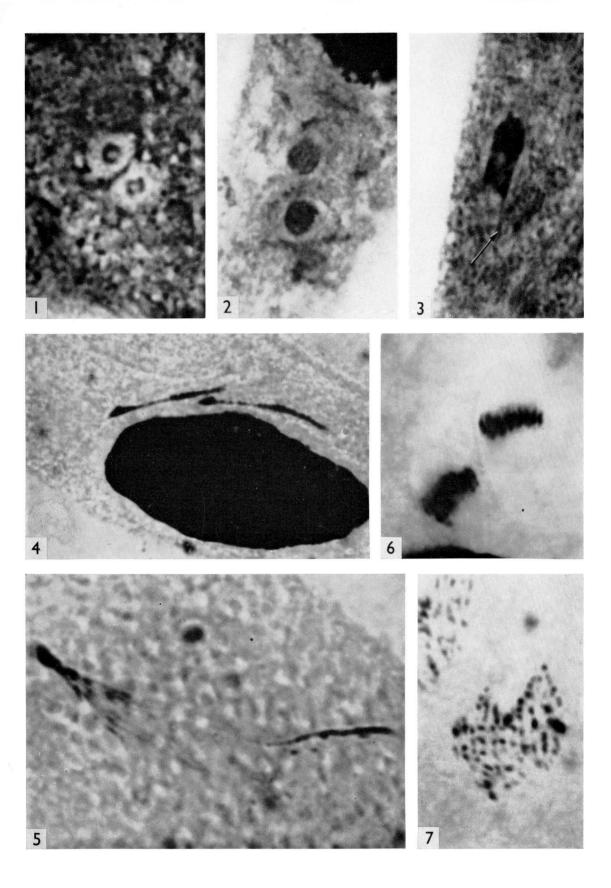

PLATE 45　Nuclear exchange at conjugation and macronuclear anlagen

These figures of whole-mounted paramecia stained with Azure-A after hydrolysis are reproduced by courtesy of Dr K. W. Jones.

Fig. 1　Light micrograph × 1500. Two paramecia are in conjugation, at stage 8 of Diagram 12. There are eight micronuclei in each conjugant resulting from two meiotic or reduction divisions. One micronucleus (arrowed) in each conjugant is seen in the paroral region and these nuclei will not degenerate. The macronucleus in each conjugant has an irregular outline, in contrast with its oval shape at the earlier stage in Plate 44, fig. 4.

Fig. 2　Light micrograph × 2200. Reciprocal nuclear exchange (as at stage 10 of Diagram 12) is taking place between gamete nuclei in the paroral region. The two nuclei shown in the paroral region appear to be migratory nuclei and the stationary nuclei are not in this focal plane. There are some degenerating nuclear fragments in the surrounding cytoplasm. There are twisted elongated fragments of the macronucleus in the upper part of the figure.

Fig. 3　Light micrograph × 1200. One exconjugant is shown at stage 14 (Diagram 12). One postconjugal mitosis of the synkaryon has taken place but the two resulting micronuclei are not shown in this photograph. The scattered fragments of the macronucleus are clearly shown and the oral space, from which macronuclear fragments are excluded, is obvious (arrow). Jones (1956) considered that the oral space played a key rôle in the differentiation of the macronucleus.

Fig. 4　Light micrograph × 2200. This squashed preparation was fixed before the first exconjugal fission, at stage 16 of Diagram 12, and shows one of the macronuclear anlage which has already grown considerably in size since the second micronuclear mitosis. The macronuclear anlage contains a number of vacuolar regions each with an intensely stained granule which contains DNA. Surrounding the macronuclear anlage there are a number of densely stained fragments of the old macronucleus.

172

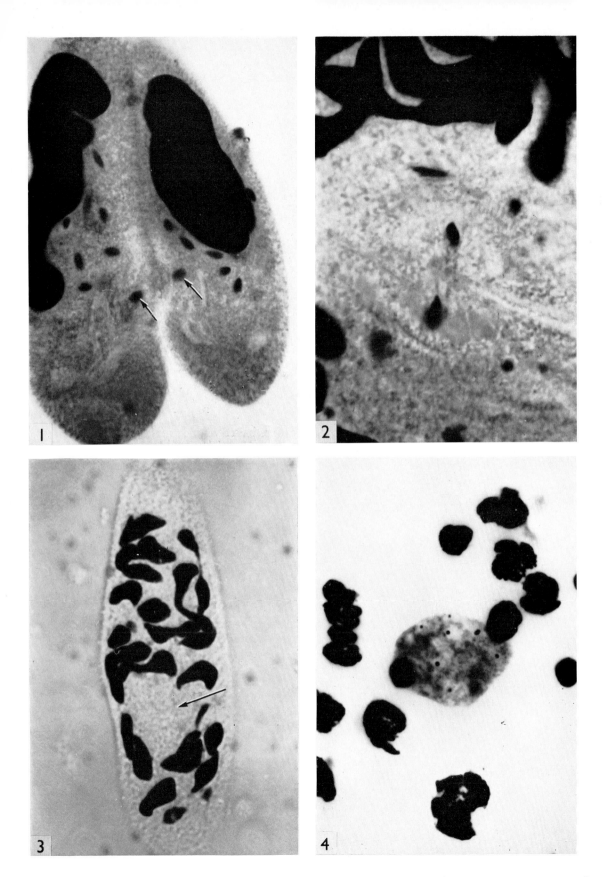

PLATE 46   Pairing at conjugation

FIG. 1   Electron micrograph ×16,000. This moderately low-power electron micrograph shows a section of two paired paramecia during conjugation. The paired ventral surfaces are sectioned down the middle of the picture from top to bottom. Although deciliation has occurred to allow the animals to pair along their ventral surfaces, cilia still can be seen along the non-paired surfaces. Many mature trichocyst bodies (mt) are present and although some of these are oriented so that their tips are below non-paired regions of the pellicle, no trichocysts can be seen oriented below the paired ventral surface. The ventral surfaces at the bottom of the figure are at a more advanced stage of pairing and pore formation than the paired surface towards the tip of the figure where deciliation is incomplete.

FIG. 2   Electron micrograph ×46,000. Part of the paired ventral surfaces are shown from a section adjacent to Fig. 1 and corresponding to the lower part of the ventral surfaces as shown in Fig. 1. Two basal granules (b) are partly shown but no cilia are present. Three cytoplasmic bridges or pores can be seen where cytoplasmic ridges have come into contact with each other. The pore to the left of the figure is closed by the plasma membranes of the conjugants.

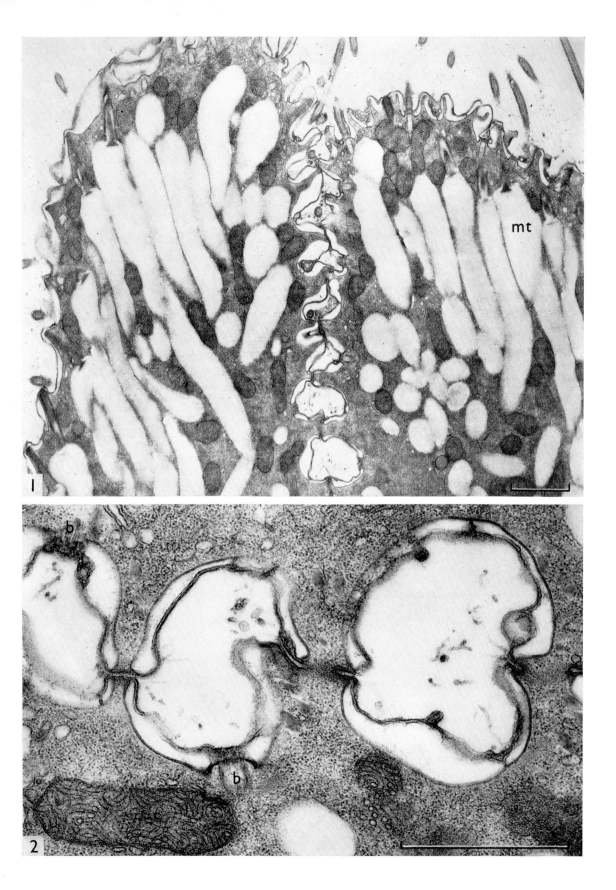

PLATE 47   Deciliation and the formation of pores

FIG. 1   Electron micrograph ×26,000. The section shows the paired ventral surfaces of two conjugants. Pairing is at quite an advanced stage because deciliation has taken place leaving the basal granules (b); the structures of the pellicle are flattened. One cytoplasmic pore has formed (arrowed) where a cytoplasmic ridge between two adjacent rows of cilia is opposite a similar ridge from the other animal. The ridges can be identified in the upper conjugant because of the presence of sectioned kinetodesmal fibrils (k). In other regions the ridges of the upper conjugant do not position themselves opposite the corresponding ridges in the lower conjugant and no pores have formed.

FIG. 2   Electron micrograph ×40,000. The section shows the paired ventral surfaces of conjugants and is similar to Fig. 1, except that the pore (arrowed) is closed by the opposed plasma membranes of the two conjugants.

FIG. 3   Electron micrograph ×45,000. This is a detail of a longitudinal section of a basal granule (b) from each conjugant, immediately opposite each other. Deciliation has occurred, but there is a short stump of a cilium joined to the lower basal granule but the upper basal granule has no stump. The strands of material between the conjugants most probably represent the breakdown products of cilia.

FIG. 4   Electron micrograph ×40,000. A detail of a longitudinal section through the mid-line of a basal granule (b) in the upper conjugant shows only a flattened surface where the cilium had once been. There is a pore to the left of the figure.

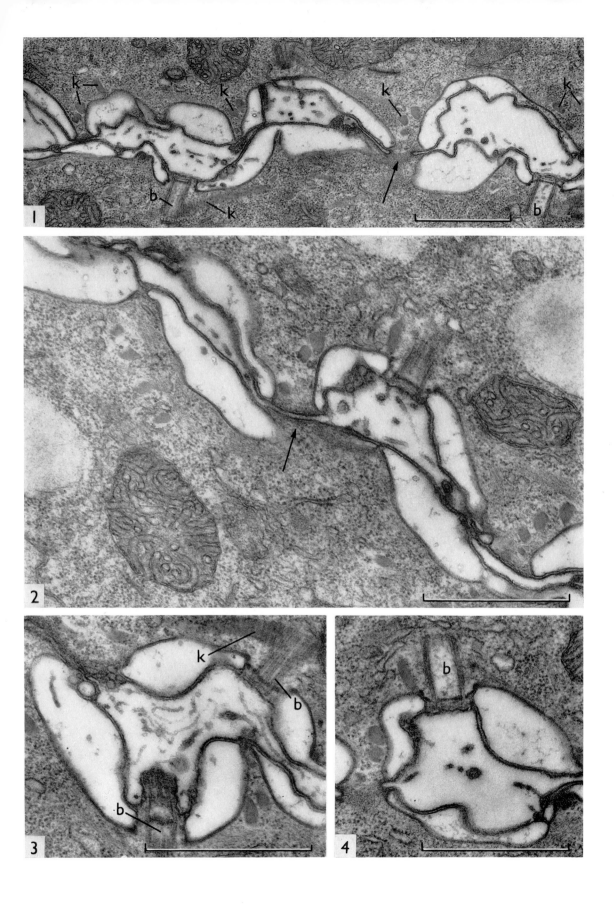

PLATE 48   Cytoplasmic pores at conjugation

FIG. 1   Electron micrograph × 48,000. The section shows paired ventral surfaces from animals in conjugation. The pellicle pattern in the two conjugants is matched and a cytoplasmic bridge or pore (arrowed) is shown at an early stage of its development in a region between cytoplasmic ridges of the two conjugants. Kinetodesmal fibrils (k) and basal granules (b) may be distinguished and the presence of a cilium between the conjugants to the right of the figure suggests that pairing is not at an advanced stage. The pore is still closed by opposed plasma membranes which are separated by a uniform distance of about 30 mμ.

FIG. 2   Electron micrograph × 60,000. This section is similar to that in Fig. 1 except that the cytoplasmic bridge or pore between the two conjugants is no longer closed by plasma membranes; instead the plasma membranes of the two conjugants have joined on either side of the pore. Granular threads (arrowed) are seen inside the pore and these have been described by Schneider (1963). This granular material is probably the breakdown product of membranes of the pellicle. In the paroral region of the conjugants, the migrating or male gamete nuclei pass through a similar cytoplasmic pore to give reciprocal fertilisation (Vivier, 1965).

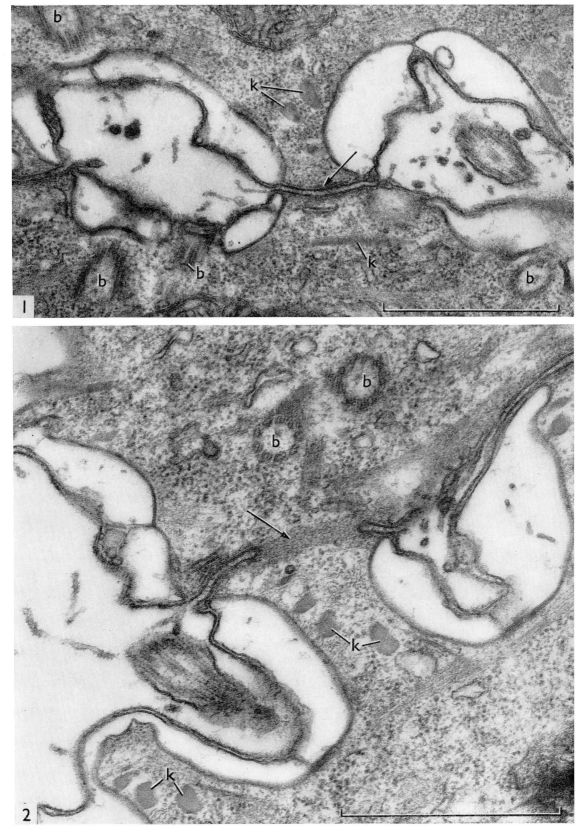

PLATE 49   Development of macronuclear anlagen

FIG. 1   Electron micrograph ×8000. The section shows a macronuclear anlage soon after the second postconjugal mitosis (stage 15 of Diagram 12). This macronuclear anlage is about 7 μ in diameter and can be distinguished from a micronucleus by its large size. It appears to contain no dense bodies but only fine fibrillar material of low electron density. Part of a fragment from the old macronucleus can be seen in the bottom right corner of the figure.

FIG. 2   Electron micrograph ×8200. The section shows a macronuclear anlage at a slightly later stage than that in Fig. 1. One large dense body of 2 μ in diameter can be seen and there are a few smaller dense bodies which are 0.1 to 0.4 μ in diameter.

FIG. 3   Electron micrograph ×6000. The two macronuclear anlagen of a paramecium after conjugation are shown separated by a layer of cytoplasm less than 0.5 μ in width (arrowed) between the nuclear membranes. The anlagen are about 10 μ in diameter and contain a few scattered dense bodies as in Fig. 2.

FIG. 4   Light micrograph ×1170. This section which is stained with toluidine blue shows the paired macronuclear anlagen (arrowed) at a stage similar to that in Fig. 3.

FIG. 5   Electron micrograph ×7500. The section shows part of one macronuclear anlage at a later stage than that shown in Figs. 3 and 4 but before the first postconjugal fission. The dense material within the anlage now appears to be in the form of sponge-like aggregates.

FIG. 6   Electron micrograph ×9300. This section is from a paramecium fixed between the first and second postconjugal fissions and shows part of the new macronucleus which developed from the macronuclear anlage. Old macronuclear fragments are still present at the top and the bottom of the figure. The new macronucleus is easily recognisable because the many small dense bodies scattered through it are appreciably smaller than the small bodies in the old macronuclear fragments.

FIG. 7   Electron micrograph ×11,000. This section was from a paramecium fixed at a similar stage to Fig. 6. In addition to the structures shown in Fig. 6, there is a region of irregular electron-dense material of considerable size (lower half of the figure) which when dispersed probably forms the large bodies of the new macronucleus.

180

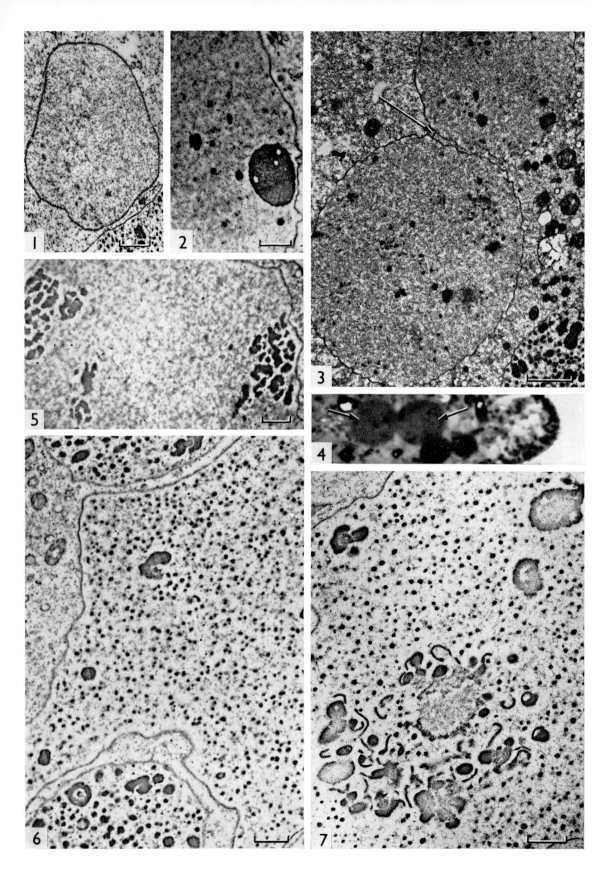

PLATE 50   Light microscopy of endosymbionts

The figures of this plate show different kinds of bacteria-like symbionts, or particles which live in the cytoplasm (in one case in the macronucleus) of various stocks of *P. aurelia* and in many cases make the host a killer for other stocks. All of the light micrographs are from material sectioned at 1 μ; they were stained with toluidine blue.

FIG. 1   Light micrograph × 1270. The longitudinal section shows many kappa particles (stock 7) in the endoplasm where they can be seen as dark dots. A dark diffuse area in the central region of the section is the macronucleus cut almost at grazing incidence. The gullet (g) is seen in section. Mature trichocysts which appear white are near the surface of the paramecium.

FIG. 2   Light micrograph × 1350. The longitudinal section is cut near the surface at the level of the trichocysts. Lambda particles (stock 239) appear as dark rod-like bodies scattered throughout the whole area of the section. The mature trichocysts appear white in the section. The lambda particles are considerably larger than the kappa in Fig. 1.

FIG. 3   Light micrograph × 2250. This section shows gamma particles (stock 565) in the endoplasm. The arrows indicate some of the clearly shown particles which are small, darkly stained and characteristically paired.

FIG. 4   Light micrograph × 1750. The longitudinal section through a specimen of *P. aurelia*, stock 562, shows elongated dark and curved alpha particles within the macronucleus (ma). The normal macronucleus without symbionts in Fig. 5 should be compared with this macronucleus. Some alpha particles also occasionally occur in the cytoplasm together with a kind of kappa particle which is specific for that stock.

FIG. 5   Light micrograph × 1200. The obliquely sectioned paramecium, stock 540, shows darkly stained clumps of mu particles (arrowed) in the endoplasm below the level of the trichocysts.

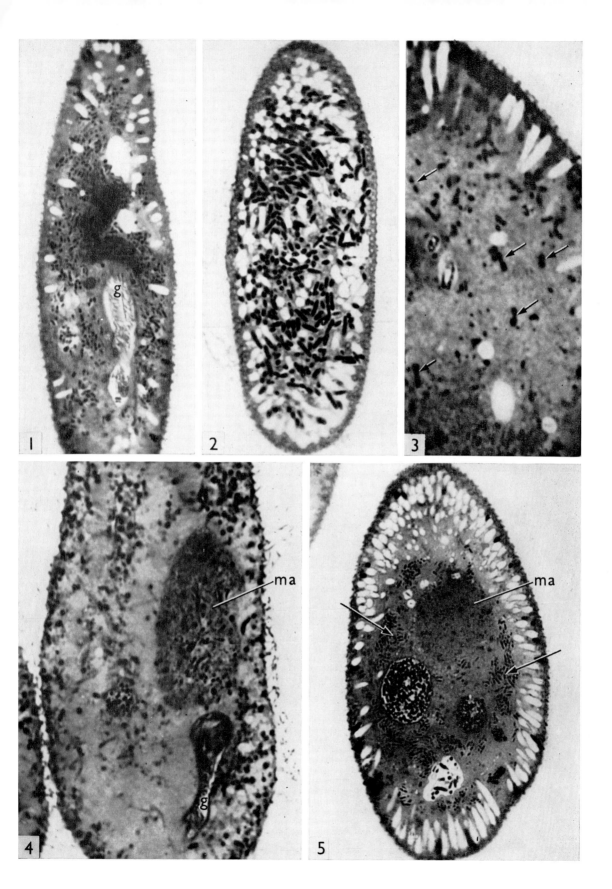

PLATE 51   Kappa particles

Figs. 1 to 3 which are reproduced by courtesy of Dr J. R. Preer are squash preparations from unfixed specimens of *P. aurelia*, and were photographed by bright phase contrast microscopy. This gives a dark background and objects appear lighter in proportion to their refractive index and thickness.

Fig. 1   Light micrograph × 5000. Kappa particles isolated from a paramecium of stock 51 are shown as grey rods. Dividing forms appear with a constriction. Some bright refractile bodies can be seen within particles (arrowed). Other refractile bodies outside the particles are probably lipid drops from the cytoplasm of the paramecium.

Fig. 2   Light micrograph × 5000. Kappa particles isolated from stock 7 appear as grey rods or spindle-shaped objects. Bright refractile bodies (arrowed) can be seen within some particles. Because the refractile body is a hollow cylinder it may appear either rectangular or as a bright disc with a darker centre according to the direction in which it is viewed. Refractile lipid drops are present.

Fig. 3   Light micrograph × 5000. This shows a number of bright refractile bodies which have been isolated from kappa particles (stock 7).

Fig. 4   Electron micrograph × 32,000. The section shows the dividing form or N type of kappa particle (stock 51). It has a constriction like the fission furrow of a bacterium in division. To the left of the figure is a more rounded form with a refractile body (rb).

Fig. 5   Electron micrograph × 40,000. A kappa particle of type B (stock 51) containing a refractile body (rb) in its cytoplasm is shown in section. The refractile body is in the form of a ribbon wound in a cylindrical coil (Diagram 14) and eight turns of the coil are seen in longitudinal section. The rupture of the cell wall of the kappa particle to the top of the figure is probably an artefact.

Fig. 6   Electron micrograph × 40,000. A similar kappa particle to that in Fig. 5 has been sectioned in a plane perpendicular to the axis of the cylindrical refractile body (rb). The fine threads in the clearer areas, here and in Fig. 5, are probably DNA.

Fig. 7   Electron micrograph × 60,000. A longitudinal section through a kappa particle (stock 7) shows a number of dense, rounded, virus-like bodies (vb) in the cytoplasm near the axis of the refractile body. The refractile body also has a sheath (arrowed) which projects from it into the granular cytoplasm of the particle. The particle has a cell wall outside its plasma membrane. This figure is reproduced by courtesy of Dr J. R. Preer and Dr A. Jurand.

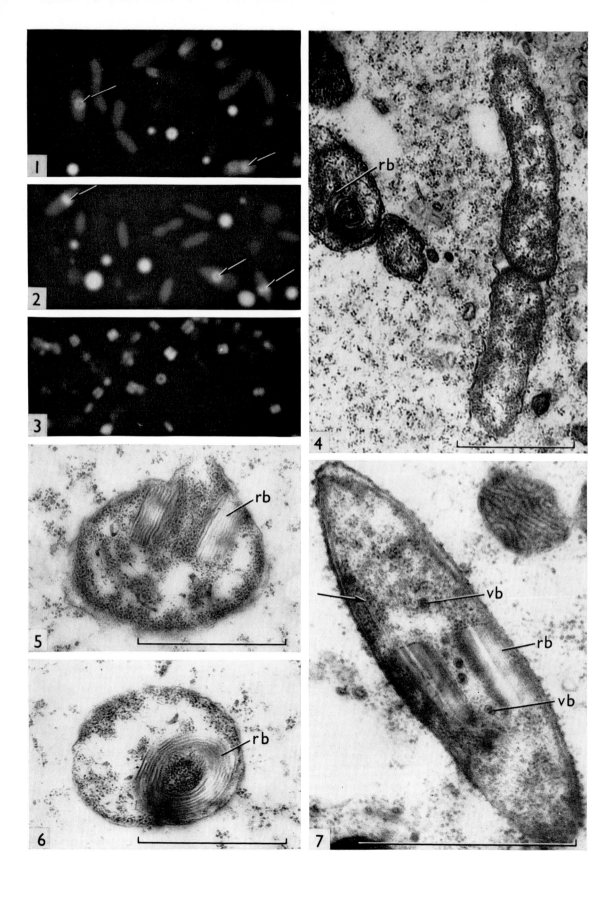

PLATE 52 Lambda particles

FIG. 1 Electron micrograph × 26,000. A negatively stained preparation of a lambda particle isolated from *P. aurelia* (stock 299) shows the particle to be rod-shaped with bacterial flagella (fl) which are 25 mµ thick and up to 5 µ in length.

FIG. 2 Electron micrograph × 200,000. This detail is from a negatively stained preparation and shows the ultrastructure of the bacterial flagella from lambda particles (stock 239). The ultrastructural units seem to be dense granules of about 8 mµ in diameter which are arranged in longitudinal rows.

FIG. 3 Electron micrograph × 45,000. This section through a lambda particle within a paramecium (stock 299) shows the particle to be in a vacuole in which bacterial flagella (fl) can be seen. Near the bottom of the figure the section includes the surface of the particle and some of the apparent granular structure is due to projections on the outer surface of the lambda cell wall. Some flagella are seen in cross-section.

FIG. 4 Electron micrograph × 29,000. A longitudinal section of a lambda particle (stock 229) shows where a fission furrow has formed at its equator.

FIG. 5 Electron micrograph × 60,000. The detail of a lambda particle (stock 239) shows its granular cytoplasm which is enclosed by a plasma membrane outside which is the cell wall with its outer projections. There are also clearer areas in the cytoplasm.

FIG. 6 Electron micrograph × 40,000. An obliquely sectioned lambda particle (stock 299) is shown with its flagella (fl) in the vacuole which has formed round the particle in the cytoplasm of the paramecium. Within the granular cytoplasm of the particle there are scattered clear areas with a few fine fibrils about 2.5 mµ in diameter. These are probably bacterial DNA.

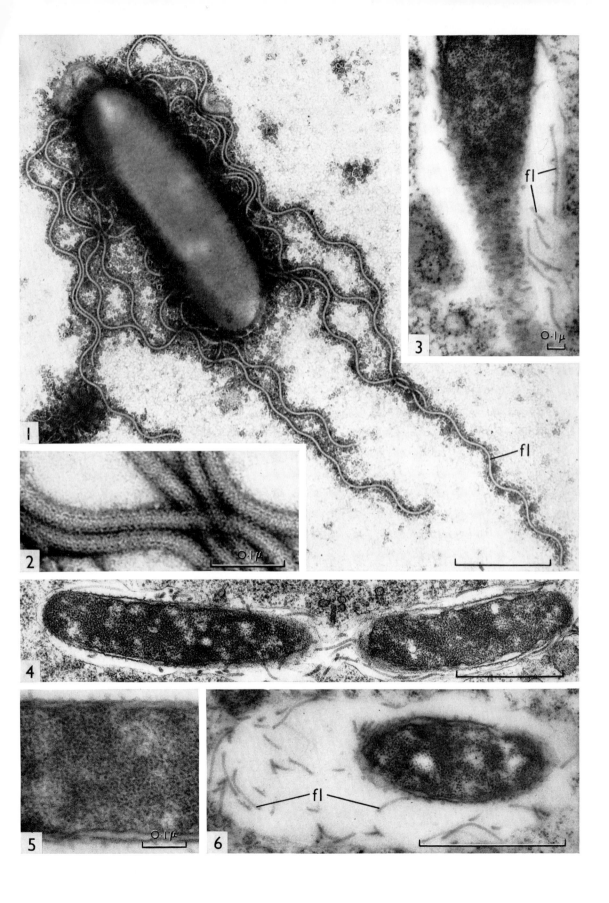

PLATE 53　Particles of stock 562

FIG. 1　Electron micrograph ×40,000. The section shows a kappa particle in the endoplasm of *P. aurelia* (stock 562). There is a refractile body (arrowed) and some dense virus-like bodies (vb) in the cytoplasm of the particle. This refractile body consists of a coiled strip which is narrower than that of stock 7 (Plate 51, fig. 7) and which has more turns than that of stock 51 (Plate 51, fig. 5).

FIG. 2　Electron micrograph ×15,000. The section shows part of the macronucleus which in stock 562 contains symbionts called alpha particles. The particles are about 6 μ long and curved but only a small part of any particle is seen in the section. The particles all have a bacterial cell wall which allows them to be distinguished from the large and small dense bodies of the macronucleus.

FIG. 3　Electron micrograph ×60,000. This detail shows a considerable length of an alpha particle in longitudinal section within a macronucleus. The granules in the cytoplasm of the alpha particle are presumably bacterial ribosomes.

FIG. 4　Electron micrograph ×45,000. This detail shows an alpha particle in longitudinal section towards one end of the particle. There are three distinct kinds of cytoplasm in the particle: a clear zone near the end of the particle (at the top of the figure), a dense finely granular zone and a zone of larger, less closely packed particles resembling the cytoplasm shown in Fig. 3. An alpha particle is seen in cross-section in the bottom left corner of the figure.

FIG. 5　Electron micrograph ×120,000. This detail shows part of an alpha particle in longitudinal section, to the left of the figure. The granular cytoplasm is bounded by a plasma membrane (pl) outside which the bacterial cell wall (arrowed) can be seen. At the top centre of the picture there is part of an alpha particle which has been sectioned transversely at the level of the dense finely granular zone of its cytoplasm.

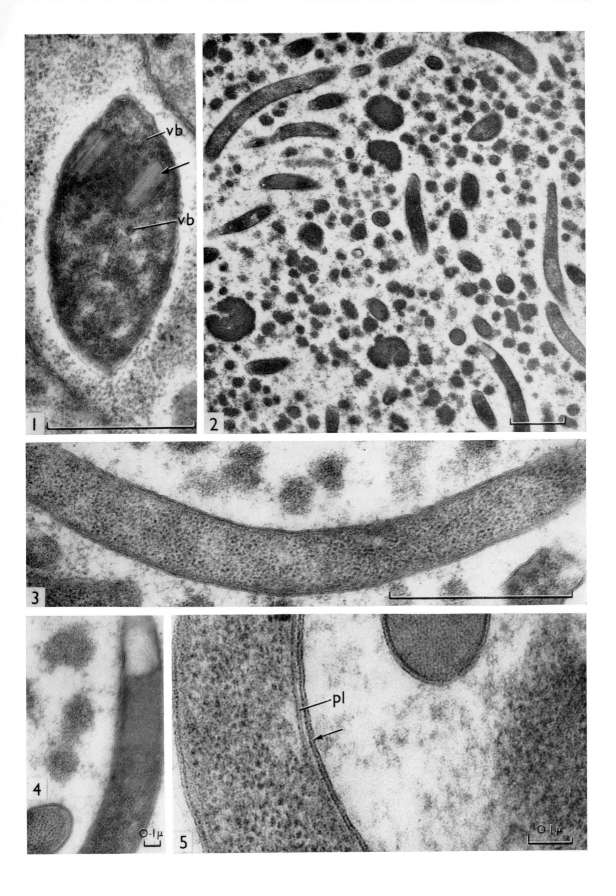

PLATE 54   Gamma and mate-killer particles

FIG. 1   Electron micrograph × 80,000. A gamma particle is shown sectioned within the cytoplasm of *P. aurelia* (stock 565). Gamma particles typically occur in pairs as shown here. In this case the particles seem to be surrounded by dense material enclosed by a unit membrane.

FIG. 2   Electron micrograph × 60,000. A gamma particle (from stock 565), is shown within a vesicle of rough endoplasmic reticulum to the outer surface of which ribosomes (arrowed) are bound. The cytoplasm within the gamma particle also has granules of the size of ribosomes.

FIG. 3   Electron micrograph × 80,000. A pair of gamma particles are shown within a vesicle of endoplasmic reticulum. The upper particle of the pair which is clearly shown has a bacterial cell wall outside the plasma membrane and dense central inclusion (arrowed) within the granular cytoplasm.

FIG. 4   Electron micrograph × 40,000. The section shows a longitudinally sectioned mate-killer or mu particle surrounded by some more mu particles in transverse section. The particles are from *P. aurelia* (stock 540). The mu particle has a bacterial cell wall and a plasma membrane while its cytoplasm contains both granules resembling ribosomes and fine fibrils uniformly distributed within the particle.

FIG. 5   Electron micrograph × 30,000. The section shows a mate-killer or mu particle from stock 540. The mu particle which is in longitudinal section has a transverse fission furrow across its central region. The mu particle is surrounded by a zone of low density material from which the cytoplasm of the paramecium is excluded. This appears to be capsular material of the mu particle.

FIG. 6   Electron micrograph × 26,000. This figure shows a longitudinal section of a mu particle which is at least 7 μ long.

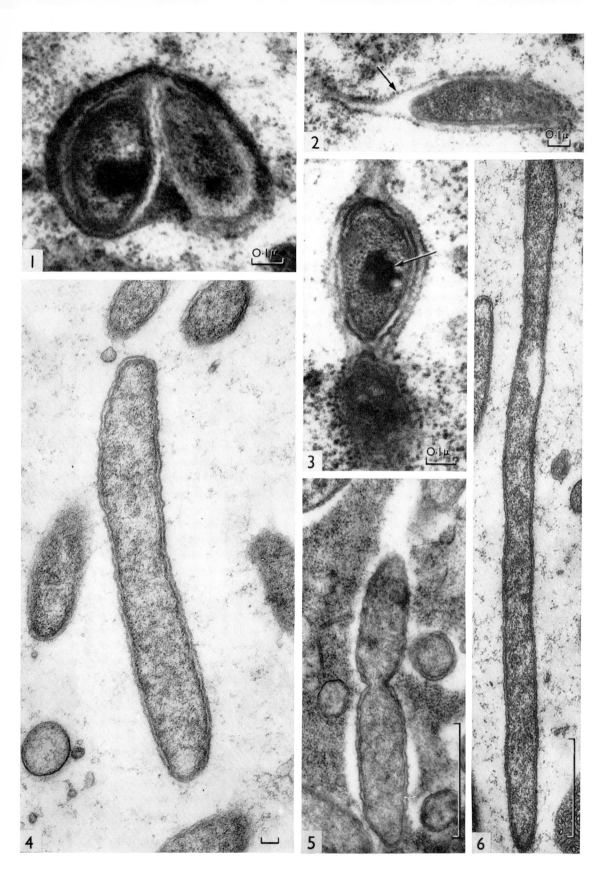

APPENDIX I   Material and methodological details

All the illustrations in this book are of *Paramecium aurelia*. Except for Plates 44 and 45 and the material used to illustrate the chapter on symbiotic inclusions, all the animals used by us were from syngen 1, stock 60.

The *Paramecium* cultures were grown and maintained in an aqueous infusion of baked lettuce, which had been inoculated with *Aerobacter aerogenes* at least 24 h previously and then adjusted to pH 7.0 by the addition of N sodium hydroxide solution.

When the animals were required for observation, the culture was first filtered through a layer of absorbent cotton wool to remove debris. Before fixation the animals were concentrated into a very small volume of liquid by filtering one litre of the culture medium through a Berkefeld candle, using the reduced pressure provided by a water pump. In this way the animals were concentrated into about 10 to 20 ml of medium which was then centrifuged at 900 r.p.m. in tubes of 15 ml capacity so that the animals were left in a dense aqueous pellet. Recently it has been found possible to dispense with the initial concentration procedure and to use instead one centrifugation with an oil-testing centrifuge and tubes of 100 ml capacity which taper towards their centrifugal ends. The supernatant fluid is decanted from the material in the pellet and the fixative is then added to the centrifuge tube which is shaken so as to disperse the animals in the fixative.

The silver-line method of Klein (1926) was used for light microscopy of the pellicle (Plate 2 and Plate 15, fig. 2). For nuclear staining of whole-mounted paramecia, Dr K. Jones used the Azure-A method of DeLamater (1961) (see Plates 44 and 45), which is based on the same principle as the better-known Feulgen method. Jurand and Bomford (1965) gave details of a paraffin wax embedding procedure which is suitable for *Paramecium*.

Material which was embedded in Araldite for electron microscopy gave excellent results by light microscopy after sectioning at 1 μ and staining with a 0.5 per cent aqueous solution of toluidine blue which was made alkaline by the addition of 1 per cent borax (Plate 41, figs. 3, 4, 6 and 7; Plate 50 and other figures). For electron microscopy, fixation in 1 per cent buffered osmium tetroxide was used routinely. The fixative was prepared according to the following formula:

Sodium barbitone	0.167 g
Sodium acetate	0.11 g
0.1 N Hydrochloric acid	11.4 ml
Osmium tetroxide	1.0 g
Sucrose	1.27 g
Calcium chloride (dry)	0.03 g
Distilled water, make up to 100 ml	

The addition of the calcium chloride in the above formula is responsible for a dramatic improvement in the preservation of fine granular constituents of the cytoplasm as well as of membranous structures. Some fixations were made without calcium chloride and although the general preservation of structure was inferior, certain fine fibrillar elements round the gullet were easier to observe in these cases because they were no longer obscured by other cytoplasmic constituents.

For *Paramecium* it has often been found an advantage to use the fixative at about 37° C, instead of in the cold. Although it is normally recommended for electron microscopy, the use of a cold fixative inevitably results in the discharge of the trichocysts and in this case the shafts expand inwards through the cytoplasm. The warm fixative allows the trichocysts to be fixed without discharge, perhaps because of the greater diffusion rate and reactivity of the osmium tetroxide at the higher temperature. The paramecia were in the fixative for 30 min and the tubes were shaken every 3 to 5 min during fixation to avoid any aggregation of the material.

Another method of fixation, modified after Afzelius (1962), has been found particularly suitable for the preservation of the various symbionts in *P. aurelia*. Carbon tetrachloride (0.3 ml) was used to dissolve osmium tetroxide (0.1 g) and the resultant solution was added to a concentrated suspension of paramecia in the culture medium (about 0.7 ml) in a test-tube. The tube was shaken gently at frequent intervals during a period of 20 min at room temperature. Because carbon tetrachloride and the culture medium do not mix, the fixation takes place in the drops of culture medium into which osmium tetroxide freely penetrates from the carbon tetrachloride. After fixation, 70 per cent ethanol was slowly added and the mixture was shaken gently until the two layers united to form one phase. The mixture was then centrifuged, the supernatant was discarded and replaced by 70 per cent ethanol.

Dehydration was usually carried out through a series of 35 per cent, 70 per cent, 95 per cent and 100 per cent ethanol and each alcohol was filtered before use. The 35 per cent alcohol was used only for 1 to 2 min as a rinse to remove the residual fixative and this step was omitted after fixation with osmium tetroxide in carbon tetrachloride. The next two higher alcohols were used for 10 min each. There were three changes of absolute alcohol each lasting for 10 min. Before each alcohol was added, the material was centrifuged in a 15 ml centrifuge tube at 900 r.p.m. for 30 sec and the supernatant liquid was carefully decanted off.

After the last change of absolute alcohol the material was embedded in Araldite as recommended by Glauert and Glauert (1958). Araldite is one of the epoxy-resins and at room temperature it has a very viscous consistency which decreases with increasing temperature. After the absolute alcohol had been decanted off, the Araldite at 50° C was poured into the centrifuge tube to four-fifths of its capacity. At this stage the paramecia usually rose to the surface of the Araldite, due to the lower specific gravity of the residual alcohol, and mixing occurred. The centrifuge tube was warmed to 60° C in a water bath and the contents were stirred by making several inversions of the tube. After 1 h the tube had cooled to room temperature. It was then warmed again to 60° C, centrifuged while still warm at about 3000 r.p.m. for 10 min and the supernatant was decanted off. The pellet containing the paramecia was then transferred to a solid watch glass where fresh Araldite was added. A rotary shaker (Jurand and Ireland, 1965) was then used to stir the cells for 45 min at a temperature of about 45° C. The Araldite suspension of paramecia was then transferred to gelatin capsules using a wide-mouthed pipette and the capsules were placed at the bottom of empty centrifuge tubes, warmed to 60° C and then centrifuged at 3000 r.p.m. for 10 min. In this way the paramecia were concentrated at the bottom of the capsules in which the polymerisation was allowed to take place.

In certain cases individual paramecia were fixed and embedded in separate capsules. The procedure was then similar to that outlined above but the animals were transferred through the alcohols using a micropipette and a glass needle was used for transference in Araldite. No centrifugation was necessary. Gelatin capsules were filled to five-sixths of their volume with partially polymerised Araldite and the animals, in less viscous Araldite, were then added to the top of the capsule. As described by Jurand (1965) the animals were oriented according to the required plane of sectioning.

The ultra-thin sections, silver or light gold in colour by reflected light, were mounted on collodion–carbon coated grids and stained by flotation on a 1 per cent solution of potassium permanganate containing 2.5 per cent uranyl acetate for 15 min. Excess stain was removed by floating the grids on distilled water, then on 0.25 per cent citric acid for 30 sec, followed by three changes of distilled water.

APPENDIX II Synonymous morphological terms and phrases

Terms used in the text are quoted first and are followed by an explanatory phrase or by some equivalent terms which are used by other authors. Only those terms are listed here which are liable to confusion.

TERMS USED	SYNONYMS
Alveolus (pl. alveoli)	Alveole (pl. alveoles) Cortical vesicle Pellicular corpuscule Peribasal space
Basal granule	Kinetosome Blepharoplast
Corpuscular unit of the pellicle	Kinetosomal territory Unit in a kinety Ciliary unit of the pellicle Kinetid
Cytopyge	Cytoproct Cell anus
Disc shaped body	Schlauchförmige Struktur
Endoral membrane	Endoral kinety Endoral ciliary membrane
Food vacuole	Gastriole
Granular cortical layer	Epiplasmic layer Homogenous layer Granular cytoplasmic layer
Gullet	Buccal apparatus
Kinetodesma (pl. kinetodesmata)	Kinetodesmos (pl. kinetodesma) Bundle of kinetodesmal fibrils

TERMS USED	SYNONYMS
Kinetodesmal fibril	Striated root fibril Myoneme and neuroneme (no longer in use)
Kinety	Row of basal granules plus associated kineto-desmal fibrils Ciliary meridian
Naked ribbed wall	Cytostomal region Cytopharyngeal region
Opisthe	Posterior cell after a fission
Plasma membrane	Cell (surface) membrane Outer pellicular membrane
Postciliary tubular fibril	Right radial ribbon
Postesophageal fibres	Postpharyngeal fibres Postoral fibres Schlundfasern
Proter	Anterior cell after a fission
Quadrulus	Vierermembran Membrana quadripartita
Transverse tubular fibril	Left tangential ribbon
Trichocyst sheath	Trichocyst cap
Vestibulum	Depressed circumoral region of the pellicle

References

AFZELIUS, B. A. (1962). Chemical fixatives for electron microscopy. *The Interpretation of Ultrastructure* (edit. by Harris, R. J. C.). New York: Academic Press (1–19).

AGAR, A. W. (1965). The operation of the electron microscope. *Techniques for Electron Microscopy* (edit. by Kay, D. H.). Oxford: Blackwell, Second Edition (1–42).

ALVERDES, F. (1922). Studien an Infusorien über Flimmerbewegung, Lokomotion und Reizbeantwortung. *Abh. Geb. exp. Biol.*, **3**, 1–133.

ANDERSON, T. F., PREER, J. R. Jr., PREER, L. B. and BRAY, M. (1964). Studies on killing particles from *Paramecium*: the structure of refractile bodies from kappa particles. *J. Microscopie*, **3**, 395–402.

ANDRÉ, J. and VIVIER, E. (1962). Quelques aspects ultrastructurales de l'échange micro-nucléaire lors de la conjugaison chez *Paramecium caudatum*. *J. Ultrastruct. Res.*, **6**, 390–406.

BALAMUTH, W. (1940). Regeneration in Protozoa; a problem of morphogenesis. *Q. Rev. Biol.*, **15**, 290–295.

BEALE, G. H. (1954). *The Genetics of Paramecium aurelia*. Cambridge University Press.

BEALE, G. H. and Jurand, A. (1960). Structure of the mate-killer (mu) particles in *Paramecium aurelia*. *J. gen. Microbiol.*, **23**, 243–252.

BEALE, G. H. and Jurand, A. (1966). Three different types of mate-killer (mu) particle in *Paramecium aurelia* (syngen 1). *J. Cell. Sci.*, **1**, 31–34.

BEALE, G. H., JURAND, A. and PREER, J. R. (1969). The classes of endosymbionts of *Paramecium aurelia*. *J. Cell Sci.* (in press).

BEISSON, J. and SONNEBORN, T. M. (1965). Cytoplasmic inheritance of the organisation of the cell cortex in *Paramecium aurelia*. *Proc. Nat. Acad. Sci. U.S.A.*, **53**, 275–282.

BUCHNER, P. (1953). In *Endosymbiose der Tiere mit pflanzlichen Mikroorganismen*. Basel: Birkhäuser.

BULLOCK, T. H. and HORRIDGE, G. A. (1965). In *Structure and Function in the Nervous System of Invertebrates*. San Francisco: Freeman.

CAROSSO, N. and FAVARD, P. (1965). Microtubules fusoriaux dans les micro et macro-nucleus de ciliés péritriches en division. *J. Microscopie*, **4**, 395–402.

CAROSSO, N., FAVARD, P. and GOLDFISCHER, S. (1964). Localisation, à l'échelle des ultrastructures, d'activités de phosphatases en rapport avec les processus digestifs chez un Cilié (*Campanella umbellaria*). *J. Microscopie*, **3**, 297–322.

CASLEY-SMITH, J. R. (1967). Some observations on the electron microscopy of lipids. *J. Roy micros. Soc.*, **87**, 463–473.

CHATTON, E. and LWOFF, A. (1935). La constitution primitive de la strie cilaire des infusoires. La desmodexie. *C.r. Séanc. Soc. Biol.*, **118**, 1068–1072.

CHATTON, E. and LWOFF, A. (1936). Techniques pour l'étude des protozoaires spéciale-ment de leurs structures superficielles (cinétome et argyrome). *Bull. Soc. fr. Microsc.*, **5**, 25–39.

CORLISS, J. O. (1959). An illustrated key to the higher groups of the ciliated protozoa, with definition of terms. *J. Protozool.*, **6**, 265–284.

DeLAMATER, E. D. (1951). A new cytological basis for bacterial genetics. *Cold Spring Harb. Symp. quant. Biol.*, **16**, 381–412.

DEMBOWSKI, J. (1924). Über die Bewegung von *Paramecium caudatum*. *Arch. Protistenk.*, **47**, 25–54.

DEMBOWSKI, J. (1962). *Historia Naturalna Jednego Pierwotniaka*. Warsaw: Państwowe Zaklady Wydawnictw Szkolnych, Second Edition.

DILLER, W. F. (1936). Nuclear reorganisation processes in *P. aurelia* with descriptions of autogamy and hemixis. *J. Morph.*, **59**, 11–67.

DIPPELL, R. V. (1954). A preliminary report on the chromosomal constitution of certain variety 4 races of *Paramecium aurelia*. *Caryologia*, **6**, (Suppl.), 1109–1111.

DIPPELL, R. V. (1958). The fine structure of kappa in killer stock 51 of *Paramecium aurelia*. Preliminary observations. *J. biophys. biochem. Cytol.*, **4**, 125–128.

DIPPELL, R. V. (1962). The site of silver impregnation in *Paramecium aurelia*. *J. Protozool.*, **9**, (Suppl.), 24.

DIPPELL, R. V. (1964). Perpetuation of cortical structure and pattern in *P. aurelia*. *Proc 11th Intern. Congr. Cell Biol., Excerpta med.*, **77**, 16–17.

DIPPELL, R. V. (1965). Reproduction of surface structure in *Paramecium*, in 'Progress in Protozoology. Abstracts of papers read at the Second International Conference on Protozoology'. *Int. Congress Series*, **91**, 65.

DIPPELL, R. V. (1968). The development of basal bodies in *Paramecium*. *Proc. natn. Acad. Sci. U.S.A.*, **61**, 461–468.

DIRKSEN, E. R. and CROCKER, T. T. (1965). Centriole replication in differentiating ciliated cells of mammalian respiratory epithelium. An electron microscope study. *J. Microscopie*, **5**, 629–644.

DOBELL, C. (1932). *Antony van Leeuwenhoek and his 'Little Animals'*. New York: Harcourt, Brace.

DRYL, S. and PREER, J. R. (1967). The possible mechanism of resistance of *Paramecium aurelia* to kappa toxin from stock 7 syngen 2 during autogamy conjugation and cell division. *J. Protozool.*, **14**, (Suppl) 33–34.

EGELHAAF, A. (1955). Cytologisch-entwicklungsphysiologische Untersuchungen zur Konjugation von *Paramecium bursaria*. *Arch. Protistenk.*, **100**, 447–514.

EHRET, C. F. (1960). Organelle systems and biological organisation. Structural and developmental evidence leads to a new look at our concepts of biological organisation. *Science*, **132**, 115–123.

EHRET, C. F. (1967). Paratene theory of the shapes of cells. *J. Theoret. Biol.*, **15**, 263–272.

EHRET, C. F., ALBLINGER, J. and SAVAGE, N. (1964). Developmental and ultrastructural studies of cell organelles. *Argonne Nat. Lab. Biol. Med. Res. Div. ann. rep.*, **6971**, 62–70.

EHRET, C. F. and DE HALLER, G. (1963). Origin, development and maturation of organelles and organelle systems of the cell surface in *Paramecium*. *J. Ultrastruct. Res.*, **6**, (Suppl.), 3–42.

EHRET, C. F. and POWERS, E. L. (1955). Macronuclear and nucleolar development in *Paramecium bursaria*. *Exp. Cell Res.*, **9**, 241–257.

EHRET, C. F. and POWERS, E. L. (1957). The organisation of gullet organelles in *Paramecium bursaria*. *J. Protozool.*, **4**, 55–59.

EHRET, C. F. and POWERS, E. L. (1959). The cell surface of *Paramecium*. *Int. Rev. Cytol.*, **8**, 97–133.

EHRET, C. F., SAVAGE N. and ALBLINGER, J. (1964). Patterns of segregation of structural elements during cell division. *Z. Zellforsch mikrosk. Anat.*, **64**, 129–139.

EHRET, C. F., SAVAGE, N. and SCHUSTER, F. L. (1965). The incorporation into cytoplasmic organelles of tritium given in thymidine. *Argonne Nat. Lab. Biol. Med. Res. ann. rep.*, **7136**, 205–210.

EHRET, C. F., SAVAGE, N. and SCHUSTER, F. L. (1966). Incorporation into cell organelles of tritium given in leucine and thymidine. *American Zool.*, **6**, 3.

FAURÉ-FREMIET, E., ROUILLER, C. and GOUCHERY, M. (1957). La réorganisation macronucléaire chez les Euplotes. *Expl Cell Res.*, **12**, 135–144.

FAVARD, P. and CARASSO, N. (1964). Étude de la pinocytose au niveau des vacuoles digestives de Ciliés Péritriches. *J. Microscopie* **3**, 671–696.

FERNÁNDEZ-MORÁN, H., ODA, T., BLAIR, P. V. and GREEN, D. E. (1964). A macromolecular repeating unit of mitochondrial structure and function. Correlated electron microscopic and biochemical studies of isolated mitochondria and submitochondrial particles of beef heart muscle. *J. Cell Biol.*, **22**, 63–100.

GALL, J. G. (1959). Macronuclear duplication in the ciliated protozoan *Euplotes*. *J. biophys. biochem. Cytol.*, **5**, 295–307.

GELEI, J. VON (1934). Der feinere Bau des Cytopharynx von *Paramecium* und seine systematische Bedeutung. *Arch. Protistenk.*, **82**, 331–362.

GELEI, J. VON (1937). Ein neues Fibrillensystem in Ectoplasma von *Paramecium*; zugleich ein Vergleich zwischen dem neuen und dem alten Gittersystem. *Arch. Protistenk.*, **89**, 133–162.

GELEI, J. VON (1939). Das aussere Stutzgerustsystem des Parameciumkörpers. *Arch. Protistenk.*, **92**, 245–272.

GIBBONS, I. R. (1961). The relationship between the fine structure and direction of beat in gill cilia of a lamellibranch mollusc. *J. biophys. biochem. Cytol.*, **11**, 179–205.

GIBBONS, I. R. (1963). Studies on the protein components of cilia from *Tetrahymena pyriformis*. *Proc. natn. Acad. Sci. U.S.A.*, **50**, 1002–1010.

GIBBONS, I. R. and GRIMSTONE, A. V. (1960). On flagellar structure in certain flagellates. *J. biophys. biochem. Cytol.*, **7**, 697–715.

GIBSON, I. and BEALE, G. H. (1961). The genic basis of the mate-killer trait in *Paramecium aurelia* stock 540. *Genet. Res.*, **2**, 82–91.

GILLIES, C. G. and HANSON, E. D. (1968). Morphogenesis of *Paramecium trichium*. *Acta Protozoologica*, **5** (in press).

GLAUERT, A. M. (1962). The fine structure of bacteria. *Br. med. Bull.*, **18**, 245–250.

GLAUERT, A. M. and GLAUERT, R. H. (1958). Araldite as an embedding medium for electron microscopy. *J. biophys. biochem. Cytol.*, **4**, 191–194.

GOLINSKA, K. (1963). Experimental study on rebounding from a mechanical obstacle in *Paramecium caudatum*. *Acta Protozoologica*, **1**, 113–120.

GRĘBECKI, A. and KUŹNICKI, L. (1961). Immobilisation of *Paramecium caudatum* in the chloral hydrate solutions. *Bull. Acad. pol. Sci. Cl. II Ser. Sci. biol.*, **9**, 459–462.

GRELL, K. G. (1967). Sexual reproduction in Protozoa. *Research in Protozoology* (edit. by Chent, T.). Oxford: Pergamon Press, **2**, 147–214.

GRIMSTONE, A. V. (1961). Fine-structure and morphogenesis in protozoa. *Biol. Rev.*, **36**, 464–536.

HAGGIS, G. H. (1966). *The Electron Microscope in Molecular Biology*. London: Longmans, Green.

HALLER, G. DE, EHRET, C. F. and NAEF, R. (1961). Technique d'inclusion et d'ultra-micotomie destinée a l'étude du dévelopement des organelles dans une cellule isolée. *Experientia*, **17**, 524–526.

HAMILTON, L. D. and GETTNER, M. E. (1958). Fine structure of kappa in *Paramecium aurelia*. *J. biophys. biochem. Cytol.*, **4**, 122–123.

HANSON, E. D. (1953). A new mutant kappa in variety 4, *Paramecium aurelia*. *Microb. Genet. Bull.*, **7**, 14.

HANSON, E. D. (1954). Studies on kappa-like particles in sensitives of *Paramecium aurelia*, variety 4. *Genetics*, **39**, 229–239.

HANSON, E. D. (1962). Morphogenesis and regeneration of oral structures in *Paramecium aurelia*. An analysis of intracellular development. *J. exp. Zool.*, **150**, 45–65.

HANSON, E. D. and GILLES, C. (1966). Oral structures during conjugation in *Paramecium aurelia*. *J. Protozool.*, **13**, (Suppl.), 26.

HARRIS, R. J. C. (1962). *The Interpretation of Ultrastructure*. New York: Academic Press.

HERFS, A. (1922). Die pulsierende Vakuole der Protozoen ein Schützorgan gegen Aussüssung. Studien über Anpassung der Organismen an das Leben im Süsswasser. *Arch. Protistenk.*, **44**, 227–260.

HERTWIG, R. (1889). Ueber die Conjugation der Infusorien. *Abh. bayer. Akad. Wiss.*, **17**, 151–233.

HILL, J. (1752). An History of Animals. (*Compleat Body of Natural History*, Vol. III, Pt. 1, Bk. 1, Fol. London.)

HOLTER, H. (1959). Pinocytosis. *Int. Rev. Cytol.*, **8**, 481–504.

HOLTER, H. (1963). Pinocytosis, in *Functional Biochemistry of Cell Structures* (edit. by Lindberg, O.). Oxford: Pergamon Press, 248–256.

HOWLAND, K. B. (1924). On excretion of nitrogen waste as a function of the contractile vacuole. *J. exp. Zool.*, **40**, 231–250.

HUFNAGEL, L. A. (1966). Fine structure and DNA of pellicles isolated from *Paramecium aurelia*. *Proc. 6th Int. Congr. Electron Microscopy, Kyoto*, 239–240.

HUFNAGEL, L. A. (1967). Physical and chemical studies on isolated pellicles of *Paramecium aurelia*. (Ph.D. Dissertation, University of Penn.)

INABA, F., SIGANUMA, Y. and IMAMOTO, K. (1966). Electron microscope observations on nuclear exchange during conjugation in *Paramecium multimicronucleatum*. *J. Protozool.*, **13**, (Suppl.), 27.

JACOB, J. and JURAND, A. (1965). Electron microscope studies on salivary gland cells 5. The cytoplasm of *Smittia parthenogenetica* (Chironomidae). *J. Insect Physiol.*, **11**, 1337–1343.

JAKUS, M. A. (1945). The structure and properties of the trichocysts of *Paramecium*. *J. exp. Zool.*, **100**, 457–485.

JAKUS, M. A. and HALL, C. E. (1946). Electron microscopic observations of the trichocysts and cilia of *Paramecium*. *Biol. Bull.*, **91**, 141–144.

JENNINGS, H. S. (1906). *Behaviour of Lower Organisms*. New York: Columbia University Press.

JENNINGS, H. S. (1913). The effect of conjugation in *Paramecium*. *J. exp. Zool.*, **14**, 279–391.

JOLLOS, V. (1921). Untersuchungen über Variabilität und Vererbung bei Infusorien. *Arch. Protistenk.*, **43**, 1–222.

JONES, K. W. (1956). Nuclear differentiation in *Paramecium*. (Ph.D. thesis, U. C. W. Aberystwyth.)

JURAND, A. (1961). An electron microscope study of food vacuoles in *Paramecium aurelia*. *J. Protozool.*, **8**, 125–130.

JURAND, A. (1962). The development of the notochord in chick embryos. *J. Embryol. exp. Morph.*, **10**, 602–621.

JURAND, A. (1965). Ultrastructural aspects of the early development of the fore-limb buds in the chick and the mouse. *Proc. R. Soc. B.*, **162**, 387–405.

JURAND, A., BEALE, G. H. and YOUNG, M. R. (1962). Studies on the macronucleus of *Paramecium aurelia*, I (with a note on ultraviolet micrography). *J. Protozool.*, **9**, 122–131.

JURAND, A., BEALE, G. H. and YOUNG, M. R. (1964). Studies on the macronucleus of *Paramecium aurelia*, II. Development of macronuclear anlagen. *J. Protozool.*, **11**, 491–497.

JURAND, A. and BOMFORD, R. (1965). The fine structure of the parasitic suctorian *Podophrya parameciorum*. *J. Microscopie*, **4**, 509–522.

JURAND, A., GIBSON, I. and BEALE, G. H. (1961). The action of ribonuclease on living paramecia. *Expl Cell Res.*, **26**, 598–600.

JURAND, A. and IRELAND, M. J. (1965). A slow rotary shaker for embedding in viscous media. *Stain Technol.*, **40**, 233–234.

JURAND, A. and PREER, L. B. (1968). Ultrastructure of flagellated lambda symbionts in *Paramecium aurelia*. *J. gen. Microbiol.*, **54**, 359–364.

JURAND, A., SIMÕES, L. C. and PAVAN, C. (1967). Changes in ultrastructure of salivary gland cytoplasm in *Sciara ocellaris* (Comstock, 1882) due to microsporidian infection. *J. Insect Physiol.*, **13**, 795–803.

KALMUS, H. (1931). *Paramecium, Das Pantoffeltierchen*. Jena: G. Fischer.

KANEDA, M. and HANSON, E. D. (1967). Growth and morphogenetic patterns in *Paramecium aurelia*. *J. Protozool.*, **14**, (Suppl.), 23.

KENNEDY, J. R. and BRITTINGHAM, E. (1968). Fine structure changes during chloral hydrate deciliation of *Paramecium caudatum*. *J. Ultrastruct. Res.*, **22**, 530–545.

KIMBALL, R. F. (1949). The effect of ultraviolet light upon the structure of the macronucleus of *Paramecium aurelia*. *Anat. Rec.*, **105**, 543.

KIMBALL, R. F. and BARKA, T. (1959). Quantitative cytochemical studies on *Paramecium aurelia*, II. *Expl Cell Res.*, **17**, 173–182.

KIMBALL, R. F., CASPERSSON, T. O., SVENSSON, G. and CARLSON, L. (1959). Quantitative cytochemical studies on *Paramecium aurelia*, I. Growth in total dry weight measured by the scanning interference microscope and X-ray absorption methods. *Expl Cell Res.*, **17**, 160–172.

KIMBALL, R. F. and GAITHER, N. (1955). Behaviour of nuclei at conjugation in *Paramecium aurelia*. I. Effect of incomplete chromosomes sets and competition between complete and incomplete nuclei. *Genetics*, **40**, 878–889.

KIMBALL, R. F. and PERDUE, S. W. (1962). Quantitative cytochemical studies on Paramecium. V. Autoradiographic studies of nucleic acid synthesis. *Expl Cell Res.*, **27**, 405–415.

KLEIN, B. M. (1926). Ergebnisse mit einer Silbermethode bei Ciliaten. *Arch. Protistenk.*, **56**, 243–279.

KOŚCIUSZKO, H. (1965). Karyologic and genetic investigations in syngen 1 of *Paramecium aurelia*. *Folia biol., Cracow*, **13**, 339–368.

KUŹNICKI, L. (1963). Recovery in *Paramecium caudatum* immobilised by chloral hydrate treatment. *Acta Protozool.*, **1**, 177–185.

LANSING, A. I. and LAMY, F. (1961). Fine structure of the cilia of rotifers. *J. biophys. biochem. Cytol.*, **9**, 799–812.

LEWIS, W. H. (1931). Pinocytosis. *Bull. Johns Hopkins Hosp.*, **49**, 17–27.

LUND, E. E. (1933). A correlation of the silverline and neuromotor systems of *Paramecium. Univ. Calif. Publs. Zool.*, **39**, 35–75.

LUND, E. E. (1941). The feeding mechanisms of various ciliated protozoa. *J. Morph.*, **69**, 563–571.

LWOFF, A. (1950). *Problems of Morphogenesis in Ciliates*. New York: John Wiley and Sons.

MANTON, I. (1959). Electron microscopical observations on a very small flagellate; the problem of *Chromulina pusilla* (Butcher). *J. mar. biol. Ass. U.K.*, **38**, 319.

MAST, S. O. (1947). The food vacuole in *Paramecium. Biol. Bull.*, **92**, 31–72.

MAUPAS, E. (1889). La rajeunissement karyogamique chez les ciliés. *Archs. Zool. exp. gén.*, **7**, 149–517.

METZ, C. B., PITELKA, D. R. and WESTFALL, J. C. (1953). The fibrillar systems of ciliates as revealed by the electron microscope. I. *Paramecium. Biol. Bull.*, **104**, 408–425.

MIJAKE, A. (1966). Local disappearance of cilia before the formation of holdfast union in conjugation of *Paramecium multimicronucleatum. J. Protozool.*, **13**, (Suppl.), 28.

MITROPHANOW, P. (1905). Étude sur la structure, le développement et l'explosion des trichocystes des Paramecies. *Arch. Protistenk.*, **5**, 78–91.

MOTT, M. R. (1963). Cytological localisation of antigens of *Paramecium* by ferritin-conjugated antibody and by counterstaining the resultant absorbed globulin. *J. Roy. microsc. Soc.*, **81**, 159–162.

MOTT, M. R. (1965). Electron microscopy studies on the immobilisation antigens of *Paramecium aurelia*. *J. gen. Microbiol.*, **41**, 251–261.

MUELLER, J. A. (1962). Induced physiological and morphological changes in the B particle and R body from killer paramecia. *J. Protozool.*, **9**, (Suppl.), 26.

MUELLER, J. A. (1963). Separation of kappa particles with infective activity from those with killing activity and identification of the infective particles in *Paramecium aurelia*. *Expl Cell Res.*, **30**, 492–508.

MULLER, J. (1856). Einige Beobachtungen an Infusorien. *Mber. Berl. Akad.*, 389–393.

MÜLLER, M. (1962). Studies on feeding and digestion in Protozoa. 5. Demonstration of some phosphatases and carboxylic esterases in *Paramecium multimicronucleatum* by histochemical methods. *Acta. biol. hung.*, **13**, 283–297.

MÜLLER, M., RÖHLICH, P., TÓTH, J. and TÖRÖ, I. (1963). Fine structure and enzymatic activity of protozoan food vacuoles. *Ciba Foundation Symposium: Lysosomes*, 201–216 (edit. by de Reuck, A. V. S. and Cameron, M. P.). London: Churchill.

MULLER, O. F. (1786). *Animalcula Infusoria fluviatilia et marina*. Copenhagen: Hauniae.

NANNEY, D. L. (1956). Caryonidal inheritance at conjugation in variety 4 of *Paramecium aurelia*. *J. Protozool.*, **4**, 89–95.

NANNEY, D. L. and RUDZINSKA, M. A. (1960). Protozoa. *The Cell. Biochemistry, Physiology, Morphology* (edit. by Brachet, J. and Mirsky, A. E.). New York: Academic Press, **4**, 109–150.

NOVIKOFF, A. B. (1961). Mitochondria (Chondriosome). *The Cell*, II, 299–421 (edit. by Brachet, J. and Mirsky, A. E.). New York: Academic Press.

PARDUCZ, B. (1959). Reizphysiologische Untersuchungen an Ziliaten, VIII. Ablauf der Fluchtreaktion bei allseitiger und anhaltender Reizung. *Ann. hist-nat. Mus. nat. hung.*, **51**, 227–246.

PARDUCZ, B. (1962). Studies on reactions to stimuli in ciliates IX. Ciliary coordination of right spiralling paramecia. *Ann. hist-nat. Mus. nat. hung.*, **54**, 221–230.

PEASE, D. C. (1960). *Histological techniques for electron microscopy*. New York: Academic Press.

PERRY, M. M. (1967). Identification of glycogen in thin sections of amphibian embryos. *J. Cell Sci.*, **2**, 257–264.

PETSCHENKO, B. DE (1911). *Drepanospira Mulleri*, n.g., n.sp. parasite des parameciums, contribution à l'étude de la structure des bactéries. *Arch. Protistenk.*, **22**, 248–298.

PITELKA, D. R. (1963). *Electron-Microscopic Structure of Protozoa*. Oxford: Pergamon Press.

PITELKA, D. R. (1964). The morphology of the cytostomal area in *Paramecium*. *J. Protozool.*, **11**, (Suppl.), 14–15.

PITELKA, D. R. (1965). New observations on cortical ultrastructure in *Paramecium*. *J. Microscopie*, **4**, 373–394.

PITELKA, D. R. (1967). Fibrillar systems in flagellates and ciliates. *Research in Protozoology*, **3**, (edit. by Chen, T.). Oxford: Pergamon Press.

PITELKA, D. R. and CHILD, F. M. (1964). The locomotor apparatus of ciliates and flagellates: relations between structure and function. *Biochemistry and Physiology of Protozoa* (edit. by Hutner, S. H.). New York: Academic Press, **3**, 131–198.

PITELKA, D. R. and PARDUCZ, B. (1962). Electron microscope observations on paralysed *Paramecium*. *J. Protozool.*, **9**, (Suppl.), 6–7.

PORTER, A. M. and BLUM, J. A. (1953). A study in microtomy for electron microscopy. *Anat. Rec.*, **117**, 685–710.

PORTER, E. D. (1960). The buccal organelles in *Paramecium aurelia* during fission and conjugation with special reference to the kinetosomes. *J. Protozool.*, **7**, 211–217.

POWERS, E. L., EHRET, C. F. and ROTH, L. E. (1955). Mitochondrial structure in *Paramecium* as revealed by electron microscopy. *Biol. Bull.*, **108**, 182–195.

POWERS, E. L., EHRET, C. F., ROTH, L. E. and MINICK, O. T. (1956). The internal organisation of mitochondria. *J. biophys. biochem. Cytol.*, **2**, (Suppl.), 341–346.

PREER, J. R. (1950). Microscopically visible bodies in the cytoplasm of the 'killer' strains of *P. aurelia*. *Genetics*, **39**, 344–362.

PREER, J. R. (1967). Genetics of the Protozoa. *Research in Protozoology*, **3** (edit. by Chen, T.). Oxford: Pergamon Press.

PREER, J. R., HUFNAGEL, L. A. and PREER, L. B. (1966). Structure and behaviour of R bodies from killer paramecia. *J. Ultrastruct. Res.*, **15**, 131–143.

PREER, J. R. and JURAND, A. (1968). The relation between virus-like particles and R bodies of *Paramecium aurelia*. *Genet. Res., Camb.*, **12**, 331–340.

PREER, J. R. and PREER, L. B. (1967). Virus-like bodies in killer *Paramecium*. *Proc. natn. Acad. Sci. U.S.A.*, **58**, 1774–1781.

PREER, J. R., SIEGEL, R. W. and STARK, P. S. (1953). The relationship between kappa and paramecin in *Paramecium aurelia*. *Proc. natn. Acad. Sci. U.S.A.*, **39**, 1228–1233.

PREER, J. R. and STARK, P. (1953). Cytological observations on the cytoplasmic factor 'kappa' in *Paramecium aurelia*. *Expl Cell Res.*, **5**, 478–491.

PROVAZEK, S. (1897). Vitalfarbung mit Neutralrot an Protozoen. *Z. wiss. Zool.*, **63**, 187–194.

RAIKOV, I. B., CHEISSIN, E. M. and BUZE, E. G. (1963). A photometric study of DNA content in macro- and micronuclei in *Paramecium caudatum, Nassula ornata* and *Lozodes magnus. Acta Protozool.*, **1**, 285–300.

RANDALL, J. T. (1957). The fine structure of the protozoan *Spirostomum ambiguum. Symp. Soc. exp. Biol.*, **10**, 185–198.

RANDALL, J. T. and DISBREY, C. (1965). Evidence for the presence of DNA at basal body sites in *Tetrahymena pyriformis. Proc. Roy. Soc. B.*, **162**, 473–491.

RAO, M. V. N. and PRESCOTT, D.M. (1967). Micronuclear RNA synthesis in *Paramecium caudatum. J. Cell. Biol.*, **33**, 281–285.

ROBERTSON, J. D. (1959). The ultrastructure of cell membranes and their derivatives. *Biochem. Soc. Symp.*, **16**, 3–43.

ROBERTSON, J. D. (1960). The molecular structure and contact relationships of cell membranes. *Prog. Biophys. biophys. Chem.*, **10**, 343–418.

ROQUE, M. (1956a). L'évolution de la ciliature buccale pendant l'autogamie et la conjugaison chez *Paramecium aurelia. C.r. hebd. Séanc. Acad. Sci., Paris*, **242**, 2592–2595.

ROQUE, M. (1956b). La stomatogenèse pendant l'autogamie, la conjugaison et la division chez *Paramecium aurelia. C.r. hebd. Séanc. Acad. Sci., Paris*, **243**, 1564–1565.

ROSENBAUM, R. N. and WITTNER, M. (1962). The activity of intracytoplasmic enzymes associated with feeding and digestion in *Paramecium caudatum*. The possible relationship to neutral red granules. *Arch. Protistenk.*, **106**, 223–240.

ROTH, L. E. (1958). A filamentous component of protozoan fibrillar systems. *J. Ultrastruct. Res.*, **1**, 223–234.

ROTH, L. E. (1959). An electron microscope study of the cytology of the protozoan *Peronema trichophorum. J. Protozool.*, **6**, 107–116.

ROTH, L. E. and SHIGENAKA, Y. (1964). The structure and formation of cilia and filaments in rumen protozoa. *J. Cell Biol.*, **20**, 249–270.

ROUILLER, C. (1960). Physiological and pathological changes in mitochondrial morphology. *Int. Rev. Cytol.*, **9**, 227–292.

ROUILLER, C. and FAURÉ-FREMIET, E. (1957). Ultrastructure réticules d'une fibre squelettique chez un Cilie. *J. Ultrastruct. Res.*, **1**, 1–13.

SAGER, R. and PALADE, G. E. (1957). Structure and development in *Chlamydomonas*, I. The normal green cell. *J. biophys. biochem. Cytol.*, **3**, 463–488.

SATIR, P. (1965). Studies on cilia, II. Examination of the distal region of the ciliary shaft and the role of the filaments in motility. *J. Cell Biol.*, **26**, 805–834.

SCHNEIDER, L. (1959a). Neue Befunde über den Feinbau des Cytoplasmas von *Paramecium* nach Einbettung in Vestopal W. *Z. Zellforsch. mikrosk. Anat.*, **50**, 61–77.

SCHNEIDER, L. (1959b). Die Feinstruktur des Nephridialplasmas von *Paramecium. Zool. Anz.*, **23**, (Suppl.), 457–470.

SCHNEIDER, L. (1960a). Elektronenmikroskopische Untersuchungen über das Nephridialsystem von *Paramecium. J. Protozool.*, **7**, 75–90.

SCHNEIDER, L. (1960b). Die Auflösung und Neubildung der Zellmembran bei der Konjugation von *Paramecium. Naturwissenschaften*, **47**, 543–544.

SCHNEIDER, L. (1961a). Elektronenmikroskopische Untersuchungen über die Wirkung von Strahlen auf das Cytoplasma. I, Die Frühwirkung von Röntgenstrahlen auf das Cytoplasma von *Paramecium. Protoplasma*, **53**, 530–553.

SCHNEIDER, L. (1961b). Elektronenmikroskopische Untersuchungen über die Wirkung von Strahlung auf das Cytoplasma. II, Die Spätwirkung von Röntgenstrahlung auf das Cytoplasma von *Paramecium. Protoplasma*, **53**, 554–574.

SCHNEIDER, L. (1963). Elektronenmikroskopische Untersuchungen der Konjugation von *Paramecium*. I, Die Auflösung und Neubildung der Zellmembran bei den Konjuganten. (Zugleich ein Beitrag zur Morphogenese cytoplasmatischer Membranen.) *Protoplasma*, **56**, 109–140.

SCHNEIDER, L. (1964a). Elektronenmikroskopische Untersuchungen an den Nahrungorganellen bei *Paramecium*. II, Die Nahrungsvakuolen und die Cytopyge. *Z. Zellforsch. mikrosk. Anat.*, **62**, 225–245.

SCHNEIDER, L. (1964b). Elektronenmikroskopische Untersuchungen an den Ernahrungs organellen von *Paramecium*. I, Der Cytopharynx. *Z. Zellforsch. mikrosk. Anat.*, **62**, 198–224.

SCHNELLER, M. V. (1962). Some notes on the rapid lysis type of killing found in *P. aurelia. Amer. Zool.*, **2**, 446.,

SEDAR, A. W. and PORTER, K. R. (1955). The fine structure of cortical components of *Paramecium multimicronucleatum. J. biophys. biochem. Cytol.*, **1**, 583–604.

SEDAR, A. W. and RUDZINSKA, M. A. (1956). Mitochondria of Protozoa. *J. biophys. biochem. Cytol.*, **2**, (Suppl.), 331–336.

SIEGEL, R. W. (1953). A genetic analysis of the mate-killer trait in *Paramecium aurelia*, variety 8. *Genetics*, **38**, 550–560.

SJÖSTRAND, F. S. (1963). A new ultrastructural element of the membranes in the mitochondria and some cytoplasmic membranes. *J. Ultrastruct. Res.*, **9**, 340–361.

SJÖSTRAND, F. S. and ELFIN, L. G. T. (1962). The layered asymmetric structures of the plasma membrane in the exocrine pancreas cells of the cat. *J. Ultrastruct. Res.*, **7**, 504–534.

SMITH, J. E. (1961). Purification of kappa particles of *Paramecium aurelia* stock 51. *Amer. Zool.*, **1**, 390.

SMITH, J. E. and VAN WAGTENDONK, W. J. (1962). Chemical identification of kappa. *Fedn. Proc. Fedn. Am. Socs. exp. Biol.*, **21**, 153.

SMITH-SONNEBORN, J. and PLAUT, W. (1967). Evidence for the presence of DNA in the pellicle of *Paramecium. J. Cell Sci.*, **2**, 225–234.

SOMMERVILLE, J. and SINDEN, R. (1968). Protein synthesis by free and bound *Paramecium* ribosomes *in vivo* and *in vitro. J. Protozool.* (in press).

SONNEBORN, T. M. (1938a). Sex behaviour, sex determination and the inheritance of sex in fission and conjugation in *P. aurelia. Genetics*, **23**, 169–170.

SONNEBORN, T. M. (1938b). Mating types in *P. aurelia*: diverse conditions for mating in different stocks: occurrence, number and interrelations of the types. *Proc. Am. phil. Soc.*, **79**, 411–434.

SONNEBORN, T. M. (1938c). Mating types, toxic interactions and heredity in *P. aurelia. Science*, **88**, 503.

SONNEBORN, T. M. (1939). *Paramecium aurelia*: mating types and groups; lethal interactions; determination and inheritance. *Am. Nat.*, **73**, 390–413.

SONNEBORN, T. M. (1946). Inert nuclei: inactivity of micronuclear genes in variety 4 of *Paramecium aurelia. Genetics*, **31**, 231.

SONNEBORN, T. M. (1947). Recent advances in the genetics of *Paramecium* and *Euplotes. Adv. Genet.*, **1**, 264–358.

SONNEBORN, T. M. (1950). Methods in the general biology and genetics of *P. aurelia. J. exp. Zool.*, **113**, 87–143.

SONNEBORN, T. M. (1953). Electron micrographs of ultra thin sections of the nuclei of *P. aurelia. Microb. Genet. Bull.*, **7**, 24.

SONNEBORN, T. M. (1954). Patterns of nucleocytoplasmic integration in *Paramecium.* Proc. 9th Int. Congr. Genet. *Caryologia*, **6**, (Suppl.), 307–325.

SONNEBORN, T. M. (1957). Breeding systems, reproductive methods, and species problems in Protozoa. *The Species Problem* (edit. by Mayr, E.), *Am. Assoc. Adv. Sci. Publ.*, **50**, 155–324.

SONNEBORN, T. M. (1958). Classification of the syngens of the *P. aurelia–multimicronucleatum* complex. *J. Protozool.*, **5**, (Suppl.), 17–18.

SONNEBORN, T. M. (1959). Kappa and related particles in *Paramecium. Adv. Virus Res.*, **6**, 229–356.

SONNEBORN, T. M. (1963). Does preformed cell structure play an essential role in cell heredity? *The Nature of Biological Diversity* (edit. by Allen, J. M.). McGraw-Hill, 165–221.

SONNEBORN, T. M. (1965). The metagon: RNA cytoplasmic inheritance. *Am. Nat.*, **99,** 279–307.

SONNEBORN, T. M., MUELLER, J. A. and SCHNELLER, M. V. (1959). The classes of kappa-like particles in *Paramecium aurelia*. *Anat. Rec.*, **134,** 642.

SONNEBORN, T. M., SCHNELLER, M. V. and CRAIG, M. F. (1956). The basis of variation in phenotype of gene-controlled traits in heterozygotes of *Paramecium aurelia*. *J. Protozool.*, **3,** (Suppl.) 8.

STEVENSON, I. (1967a). Biochemical capabilities of the mu particles of *Paramecium aurelia*. *J. Protozool.*, **14,** (Suppl.), 43.

STEVENSON, I. (1967b). A method for the isolation of macronuclei from *Paramecium aurelia*. *J. Protozool.*, **14,** 412–414.

STEWART, J. M. and MUIR, A. R. (1963). The fine structure of the cortical layers in *Paramecium aurelia*. *Quart. J. miscrosc. Sci.*, **104,** 129–134.

SUYAMA, Y. and PREER, J. R. (1965). Mitochondrial DNA from protozoa. *Genetics*, **52,** 1051–1058.

VIVIER, E. (1965). Sexualité et conjugation chez la paramecie. *Ann. Fac. Sc. Clermont Ferrand*, **26,** 101–114.

VIVIER, E. (1966). Variations ultrastructurales du chondriome en relation avec le mode de vie chez des protozoaires. *Proc. 6th Int. Congr. Electron Micr., Kyoto*, 247–248.

VIVIER, E. and ANDRÉ, J. (1961a). Existence d'inclusions d'ultrastructure fibrillaire dans le macronucleus de certains souches de *Paramecium caudatum* Ehr. *C.r. hebd. Séanc. Acad. Sci., Paris*, **252,** 1848–1850.

VIVIER, E. and ANDRÉ, J. (1961b). Données structurales et ultrastructurales nouvelles sur la conjugaison de *Paramecium caudatum*. *J. Protozool.*, **8,** 416–426.

VOLKONSKY, M. (1934). L'aspect cytologique de la digestion intracellulaire. *Arch. exp. Zellforsch.*, **15,** 355–372.

WADDINGTON, C. H. and PERRY, M. M. (1962). The ultrastructure of the developing urodele notochord. *Proc. Roy. Soc. B.*, **156,** 459–482.

WEATHERBY, J. H. (1927). The function of the contractile vacuole in *Paramecium caudatum*, with special reference to the excretion of nitrogenous compounds. *Biol. Bull.*, **52,** 208–218.

WAGTENDONK, W. J. VAN (1948). The killing substance paramecin: chemical nature. *Amer. Nat.*, **82,** 60–68.

WAGTENDONK, W. J. VAN and TANGUAY, R. B. (1963). The chemical composition of lambda in *Paramecium aurelia*, stock 299. *J. gen. Microbiol.*, **33,** 395–400.

WAGTENDONK, W. J. VAN, CLARK, A. D. and GODOY, G. A. (1963). The biological status of lambda and related particles in *Paramecium aurelia. Proc. natn. Acad. Sci. U.S.A.*, **50,** 835–838.

WEISZ, P. B. (1951). An experimental analysis of morphogenesis in *Stentor Coeruleus. J. exp. Zool.*, **116,** 231–257.

WICHTERMAN, R. (1953). *The Biology of Paramecium.* New York: Blakiston.

WOHLFARTH-BOTTERMANN, K. E. (1956a). Entstehung, Feinstruktur und Vermehrung der Mitochondrien von *Paramecium. Verh. dt. zool. Ges., Hamburg,* 242–249.

WOHLFARTH-BOTTERMANN, K. E. (1956b). Protistenstudien 7. Die Feinstruktur der Mitochondrien von *Paramecium caudatum. Z. Naturf.*, **11b,** 578–581.

WOHLFARTH-BOTTERMANN, K. E. (1957). Cytologische Studien. 4, Die Entstehung, Vermehrung und Sekretabgabe der Mitochondrien von *Paramecium. Z. Naturf.,* **12b,** 164–167.

WOHLFARTH-BOTTERMANN, K. E. (1958a). Cytologische Studien, IV. Die Feinstruktur des Cytoplasmas von *Paramecium. Protoplasma,* **49,** 231–247.

WOHLFARTH-BOTTERMANN, K. E. (1958b). Elektronen-Mikroskopie. Neue Erkenntnisse zum Feinbau der Zelle. *Umschau,* **5,** 144–147.

WOHLFARTH-BOTTERMANN, K. E. and SCHNEIDER, L. (1961). Strahlenwirkungen an Mitochondrien. *Strahlentherapie,* **116,** 25–38.

WOLFE, J. (1967). Structural aspects of amitosis: a light and electron-microscope study of the isolated macronuclei of *Paramecium aurelia* and *Tetrahymena pyriformis. Chromosoma,* **23,** 59–79.

WOODWARD, J., GELBER, B. and SWIFT, H. (1961). Nucleoprotein changes during the mitotic cycle in *Paramecium aurelia. Expl Cell Res.,* **23,** 258–264.

WOODWARD, J., WOODWARD, M., GELBER, B. and SWIFT, H. (1966). Cytochemical studies of conjugation in *Paramecium aurelia. Expl Cell Res.,* **41,** 55–63.

YUSA, A. (1957). The morphology and morphogenesis of the buccal organelles in *Paramecium* with particular reference to their systematic significance. *J. Protozool.,* **4,** 128–142.

YUSA, A. (1963). An electron microscope study on regeneration of trichocysts in *Paramecium caudatum. J. Protozool.,* **10,** 253–262.

YUSA, A. (1964). Observations on atypical mitochondria in *Paramecium caudatum. J. Protozool.,* **11,** 237–239.

YUSA, A. (1965). Fine structure of developing and mature trichocysts in *Frontonia vesiculosa. J. Protozool.,* **12,** 51–60.

Index

The page numbers which are in italic refer to the page of text accompanying the plate on the adjacent page.

Jurand, Artur.
    The anatomy of *Paramecium aurelia* [by] A. Jurand and
G. G. Selman.   London, Macmillan; New York, St. Martin's P., 1969.

    xiii, 218 p. illus. 26 cm. 5/-/-   B 69–18578

    Bibliography: p. 199–212.

    1. Paramecium aurelia. 2. Ciliata—Anatomy.   I. Selman, G. G.,
joint author. II. Title.